Applied ethology 2014:

Moving on

ISAE2014

Proceedings of the 48th Congress of the International Society for Applied Ethology

29 July – 2 August 2014, Vitoria-Gasteiz, Spain

Moving on

edited by:

Inma Estevez

Xavier Manteca

Raul H. Marin

Xavier Averós

Wageningen Academic
P u b l i s h e r s

EAN: 9789086862450
e-EAN: 9789086867974
ISBN: 978-90-8686-245-0
e-ISBN: 978-90-8686-797-4
DOI: 10.3920/978-90-8686-797-4

First published, 2014

© Wageningen Academic Publishers
The Netherlands, 2014

Welcome to the 48th Congress of the ISAE

What makes science most exciting is not how much you know, but how much you can still learn, widening the possibilities of exploring new horizons. In this learning process diversity of experiences, exposure to new ideas, concepts or methodologies enrich and expand our capacity for innovation. This is why we centred our main goal for the 2014 ISAE Congress in providing the audience with a multidisciplinary and diverse forum. Our aim was not only to bring the most recent advances in applied animal behaviour and animal welfare, but also raise awareness of new interdisciplinary approaches, ideas and tools that would allow us to further advance in the study of animal behaviour and welfare. The scientific program 'MOVING ON' that you are about to experience will cover a great variety of traditional, but also many non-traditional topics such as: movement and space use, modelling and social networking, precision/smart farming, from pain to positive emotions, clinical behavioural problems, welfare in wildlife, neurobiology of behaviour and welfare, and behaviour and reproduction. The congress will feature several challenging plenary lectures that will present the state-of-the-art in applied behaviour sciences and animal welfare, free communications and posters, in addition to specialized workshops, and exciting social events. Furthermore, in this Congress you will have a chance to celebrate with us the 50th Anniversary of the publication of Ruth Harrison's book 'Animal Machines', the first book which exposed the welfare issues of the intensive farming. We sincerely hope you will greatly enjoy this special celebration and the scientific program.

We also decided to take some risk in changing partially the format of the Congress. This year we are giving you ´free choice´ on having hard copies of the proceedings, staying for the excursion depending on your ´accessibility to resources´, or by giving you a chance to promote your poster with your own ´video spot´ for the first time ever!!! As Einstein said ´Anyone who has never made a mistake has never tried anything new´, so we are trying new ideas with the sincere hope that they will work. If they do not, next Congress organizers will know what not to try again!

Our most warm welcome to Basque Country and to ISAE 2014 MOVING ON!
Ongi etorri Euskadiko!

Inma Estevez & Xavier Manteca

Acknowledgements

Congress Organising Committee

Inma Estevez (chair)	Neiker-Tecnalia & Ikerbasque
Xavier Manteca (co-chair)	Universitat Autònoma de Barcelona
Marta Alonso	Universidad de León
Marta Amat	Universitat Autònoma de Barcelona
Xavier Averos	Neiker-Tecnalia
Ina Beltrán de Heredia	Neiker-Tecnalia
Tomàs Camps	Universitat Autònoma de Barcelona
Ricardo Cepero	Universidad de Zaragoza
Guiomar Liste	Neiker-Tecnalia
Nerea Mandaluniz	Neiker Tecnalia
Eztiñe Ormaetxea	ELIKA
Roberto Ruiz	Neiker-Tecnalia
Antonio Velarde	IRTA
Raul Marin	Neiker-Tecnalia

Scientific Committee

Xavier Manteca (chair)	Universitat Autònoma de Barcelona
Inma Estevez (co-chair)	Neiker-Tecnalia & Ikerbasque
Ana Cristina Barroeta	Universitat Autònoma de Barcelona
Francisca Castro	Consejo Superior de Investigaciones Científicas.
Antonio Dalmau	IRTA
Silvia García	Universidad de Zaragoza
Jorge Palacio	Universidad de Zaragoza
Jose M. Peralta	Western University of Health Sciences, California USA
Tomas Redondo	CSIC
Deborah Temple	Universitat Autònoma de Barcelona
Miguel Angel Aparicio Tovar	Universidad de Extremadura
Arantxa Villagrá	Centro de Tecnología Animal CITA-IVIA
Morris Villaroel	Universidad Politécnica de Madrid
Jesús Martín Zuñiga	Universidad de Granada

Reviewers

Algers, B.
Amat, M.
Andersen, I.L.
Averós, X.
Barroeta, A.C.
Bartosova, J.
Blache, D.
Blokhuis, H.
Boissy, A.
Buijs, S.
Butterworth, A.
Casey, R.
Cassinello, J.
Cheng, H.W.
Corboulay, V.
Cozzi, A.
Croney, C.
Dalmau, A.
Damian, J.P.
de Jong, I.
Devillers, N.
Diaz, S.
Duncan, N.
Dwyer, C.
Earley, B.
Peralta, J.
Fàbrega, E.
Faraco, C.B.
Faucitano, L.
Ferret, A.
Galindo, F.
Gallo, C.
García-Belenguer, S.

Gispert, M.
Goddard, P.
Gonzalez, G.
González, L.
Gunnarsson, S.
Hargtung, J.
Hill, S.
Houpt, K.
Huertas, S.M.
Hultgren, J.
Jensen, M.B.
Jensen, P.
Jones, B.
Jovani, R.
Keeling, L.
Knierim, U.
Koene, P.
Leone, E.H.
Llonch, P.
Mainau, E.
Makagoon, M.
Manteca, X.
María-Levrino, G.
Martín-Zúñiga, J.
Meunier-Salaum, M.C.
Mills, D.
Minero, M.
Moe, R.
Morton, D.
Mullan, S.
Munksgaard, L.
Newberry, R.
Nicol, C.

Olsson, A.
Padros, F.
Pageat, P.
Pajor, E.
Palacio, J.
Paranhos, M.
Pluijmakers, J.
Raj, M.
Rodemburg, B.
Roque, A.
Siracusa, C.
Spoolder, H.
Sylvie, C.
Tallet, C.
Temple, D.
Thaxton, Y.V.
Tuyttens, F.
Valros, A.
Velarde, A.
Vergara, P.
Vermeer, H.
Villagrà, A.
Villalba, J.
Villaroel, M.
von Borrell, E.
von Holleben, K.
von Keyserlingk, M.
Waiblinger, S.
Weary, D.
Winckler, C.
Zanella, A.

EUROPA

CONFERENCE AND EXHIBITION CENTRE

Why Vitoria-Gasteiz?

Vitoria-Gasteiz, the capital of the Basque Country, is a beautiful medieval city, green and undiscovered by mainstream tourists. **The Europa Conference and Exhibition Centre** is a modern eco-building, ideal for events of up to 1000 people. Hope to welcome you soon!

European Green Capital 2012
Spanish Capital of Gastronomy 2014

PURINA
PRO PLAN®

PROVEN to improve cognitive function in Senior dogs

❝ WITH PRO PLAN® ANTI AGE* I'M COMMITTED TO PROVIDING A HEALTHY & HAPPY LIFE FOR MY SENIOR DOG. ❞

Karine,
Research Scientist
Creator of **PRO PLAN® ANTI AGE***
with Brillo.

Clinically PROVEN to enhance cognitive function in ageing dogs#

*patent pending #Pan et al., British J Nutr, (2010) 103, 1746-1754

Institut de Recherche en Sémiochimie et Éthologie Appliquée
Research Institute in Semiochemistry and Applied Ethology

HUMANE SOCIETY
INTERNATIONAL

Sponsors

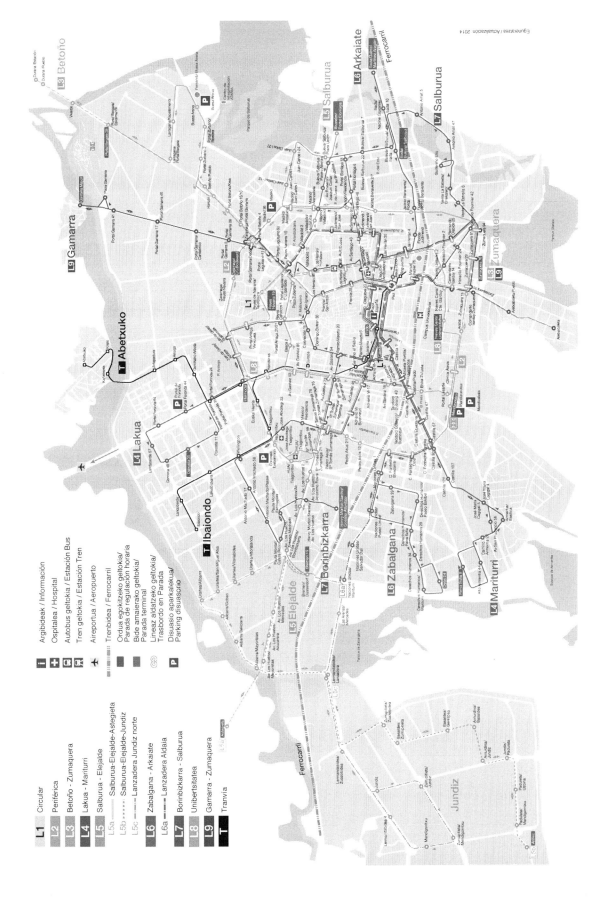

Vitoria-Gasteiz

Vitoria-Gasteiz is the capital of the Basque Country in northern Spain, with a population of 235,661 people. It is the second largest Basque city. The Spanish Gastronomy Capital prize is awarded to Vitoria-Gasteiz in 2014.

The city won the "European Green Capital 2012" award which recognized Vitoria-Gasteiz as the city in the European Union most committed to environmental conservation. Vitoria-Gasteiz is one of the European cities with the largest amount of green spaces and gardens per person (approx. 42 square meters per capita including the current extension of the Green Ring). The Basque capital offers over ten million square meters of parkland and green areas for walking, cycling, riding, bird watching and observing deer.

Sightseeing

The old quarters of Vitoria-Gasteiz, were declared a Monumental Site in 1997, still preserve their medieval layout. Three Europa Nostra Awards support the work done in the rehabilitation of areas and landmarks at the Medieval Quarters.

The Cathedral of Santa María, The Basilica of Armentia, the churches (San Pedro Apóstol, San Miguel Arcángel and San Vicente Mártir).

The palaces (The Cord House, the Palace of Bendaña, the Palace of Montehermoso, the Palace of Escoriaza-Esquibel, the Palace of Villasuso, the Palace of Diputación and the Palace of Zulueta).

The squares and open spaces (Plaza de España and the Town Hall, Plaza de la Virgen Blanca, Plaza de los Fueros, Plaza del Machete, The Arquillos).

The Green Belt The city is surrounded by a green belt consisting of six major parks, crossed by an extensive network of pedestrian-cyclist routes furnished with extensive rest areas.

Useful information

Official Language

The language of the Congress is English.

The Weather

Temperatures are moderate throughout the year. It rains more frequently in the spring and autumn and there is some snow in winter. Our summers are not excessively hot. The average temperature in summer is 18 °C and in winter 8 °C.

More information: Euskalmet and World Meteorological Organization

Local currency

The Spanish monetary unit is the Euro. Major credit cards are accepted almost everywhere. There are several currency exchange ATM at Bilbao International Airport and you will be able to exchange money in most of banks in Vitoria-Gasteiz.

Exchange rates may vary. To see current exchange rates, please visit www.oanda.com or www.x-rates.com.

Timetables

The normal working hours in offices are as follows: Mornings from 9 am to 1/2 pm and afternoons from 3 or 4 pm to 7 pm. On Saturdays, most companies and offices are closed. Official centers are generally open to the public only in the morning (from 8 am to 2 pm).

The usual meal times are:
- Breakfast: 8 to 10 am
- Lunch: 13:30 to 15:30 pm
- Dinner: 21 to 23 pm

Shops normally open from 10 AM to 1:30 pm and from 5 pm to 8 pm. Many shops open on Saturdays. Department stores and shopping centres do not generally closed at midday. Their opening times are from 10 am to 10 pm, Saturdays included.

Official time is, as in the rest of Spain, as follows:
- in the autumn-winter (GMT+1)
- in the spring-summer (GMT+2)

Electrical current

The normal electrical current is 220 V and 50 Hz. European type plugs with rounded pins are used.

Banks and savings banks

The public opening hours of banks and savings banks is from 8:30 am to 2 pm. Some branches also open in the afternoon, from 4 pm to 5:30 pm.

If you lose your banking card or in the case of robbery, these are the numbers to call:

Service telephone
VISA: 91 519 21 00
SERVIRED: 91 519 21 00
AMERICAN EXPRESS: 902 37 56 37
4B: 902 11 44 00

Passports

Citizens of the European Community and some Eastern European countries can enter the country with their National Identity Document. For citizens of other countries, a current passport or passport and visa are required as applicable. Please get in touch with the Spanish embassy or consulate in your country.

Health information

No vaccination of any kind is required to enter the country.

Useful telephone numbers:

Emergencies and 24-hour services
Emergencies 112
Local Police 092
Ertzaintza (Basque police): Police station Vitoria-Gasteiz 945 064219
Pharmacies (Information) 945 230721
Red Cross 945 222222

Venue

Palacio Europa

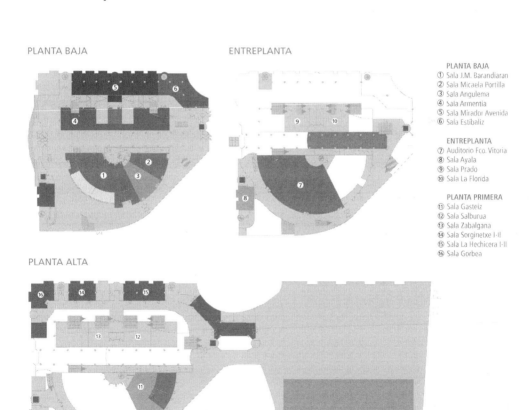

PLANTA BAJA

ENTREPLANTA

PLANTA ALTA

PLANTA BAJA
① Sala J.M. Barandiaran
② Sala Micaela Portilla
③ Sala Angulema
④ Sala Armentia
⑤ Sala Mirador Avenida
⑥ Sala Estibaliz

ENTREPLANTA
⑦ Auditorio Fco. Vitoria
⑧ Sala Ayala
⑨ Sala Prado
⑩ Sala La Florida

PLANTA PRIMERA
⑪ Sala Gasteiz
⑫ Sala Salburua
⑬ Sala Zabalgana
⑭ Sala Sorginetxe I-II
⑮ Sala La Hechicera I-II
⑯ Sala Gorbea

ISAE2014
48th Congress of the International Society for Applied Ethology

JUL29-AUG02 2014 Palacio Europa Vitoria-Gasteiz

program
www.isae2014.com

UAB — Universitat Autònoma de Barcelona · neiker

TUESDAY 29 JULY
- Registration Europa Palace
- Sat WORKSHOP 1: "Prenatal and early postnatal effects on offspring development" - J.Andersen et al.
- Sat WORKSHOP 2: "Open rearing in dairy production" - J.Lidfors et al.
- Council Meeting
- Opening Ceremony
- Reception Europa Palace

WEDNESDAY 30 JULY
- Registration Europa Palace
- Wood-Gush Memorial Lecture: Agent-based modelling: a powerful tool for applied ethology - Prof. Volker Grimm
- PLENARY: Precision livestock farming to improve animal welfare and health - Prof. Daniel Berckmans
- Coffee Break @ Poster Session (1)
- S1: Precision farming
- S2: Free papers I
- Lunch (@ conference Center- Pavilion)
- Coffee served @ Poster Session (1)
- PLENARY: New approaches for modelling and analysis of animal movement behaviour - Dr. Edward Codling
- S3: Movement patterns & space use
- S4: Free papers II
- Coffee Break
- S3 (Cont.)
- S5: Cognition
- S6: Social networking and modeling
- Wine and cheese networking event

THURSDAY 31 JULY
(Celebrating 'animal machines')
- Editorial Board ABBS Meeting (Board members)
- PLENARY: Ruth Harrison's animal machines 50 years on: still providing lessons for the future - Prof. Marian Stamp Dawkins
- PLENARY: 50 years on: are animals still machines? - Prof. Mike Mendl
- Coffee Break @ Poster Session (2)
- S7: From pain to positive emotions
- S8: Methodology in animal welfare Res.
- S9 Companion animals
- S10: Free papers III
- S11: Welfare in wildlife
- ISAE Regional Secretaries Lunch
- Lunch @ conference Center- Pavilion
- Coffee served @ Poster Session (2)
- PLENARY: Observer bias in applied ethology: can we believe what we score, if we score what we should? - Dr. Frank Tuyttens
- S10 (Cont.)
- S11 (Cont.)
- Coffee Break
- WORKSHOP 1: Methodology in animal welfare research – F.Tuyttens.
- WORKSHOP 2: Clinical behavioural problems – J.Yeates, M.Amat & T.Camps.
- WORKSHOP 3: Bioethics in wild life experimentation – A.Olsson.
- Free time* (city visit suggestions: medieval town, pintxo-pote routes...)

*Pintxo-pote: includes a glass of wine or beer and a 'tapa' (pintxo in the Basque) for about 1.20-1.5 € serve at many bars in the City on Thursdays

FRIDAY 01 AUGUST
- PLENARY: Enhancing cognitive function in rodents - Prof. Shira Knafo
- PLENARY: The architecture of sexual behaviour: lessons from the chicken - Dr. Tommaso Pizzari
- Coffee Break @ Poster Session (3)
- S12: Neurobiology of behaviour and welfare
- S13: Free papers IV
- Lunch (@ conference Center- Pavilion)
- Coffee served @ Poster Session (3)
- S14: Behaviour and reproduction
- S15: free papers V
- Closing Ceremony
- ISAE General Meeting
- Congress Banquet

SATURDAY 02 AUGUST
- Excursions

Time scale: 08:00 · 08:30 · 09:00 · 09:30 · 10:00 · 10:30 · 11:00 · 11:30 · 12:00 · 12:30 · 13:00 · 13:30 · 14:00 · 14:30 · 15:00 · 15:30 · 16:00 · 16:30 · 17:00 · 17:30 · 18:00 · 18:30 · 19:00 · 19:30 · 20:00 · 20:30 · 21:00 · 21:30 · 22:00 · 22:30 · 23:00 · 23:30 · 00:00 · 01:00

Scientific and Social Program

Tuesday July 29

09:00-13:30	Council Meeting (ISAE Council Members)
15:30-19:00	Registration Europa Palace

16:00-18:00 SATELLITE WORKSHOP 1	16:00-18:00 SATELLITE WORKSHOP 2
Prenatal and early postnatal effects on offspring development. *I. L. Andersen, R. Newberry, X. Averós, Rachel Chojnaki and Emma Baxter*	**Dam rearing in dairy production** *K. A. Zipp, J.J. Johnsen, T. Kalber, C. Mejdell, K. Bar and U. Knierim*

19:00-21:30   Opening and Reception Europa Palace

Wednesday July 30

08:00-08:30	Registration Europa Palace

8:30-10:00   Wood-Gush Memorial Lecture: Agent-based modelling: a powerful tool for applied ethology. *Volker Grimm*

10:00-10:45   PLENARY: Precision livestock farming to improve animal welfare and health. *Daniel Berckmans*

10:45-11:45	Coffee Break and Poster Session (1)

11:45-13:30 SESSION 1: Precision farming	SESSION 2: Free papers I
11:45-12:00	**Cardiac activity, behavior and cortisol release in dairy cattle during interaction with scraper robots.** *Renate Luise Doerfler, Jana Ebert, Anne Gramsch, Katharina Haas, Sylvia Kuenz, Martin Pardeller, Heike Kliem, Martin Bachmaier and Heinz Bernhardt*	**Assessment of cage space in breeding mice.** *Brianna N. Gaskill and Kathleen R. Pritchett-Corning*
12:00-12:15	**The animal-machine interface: dairy calves' adaptation to automated feeders.** *Mayumi Fujiwara, Jeffrey Rushen and Anne Marie De Passille*	**Development of an exploratory test and a contrast test for mice.** *Anette Wichman and Elin Spangenberg*
12:15-12:30	**The accuracy of the automatic recording of lying behaviour is affected by whether dairy cows are inside or outside.** *S. Mark Rutter, Norton E. Atkins and Stephanie A. Birch*	**Utilization of nesting material by laboratory hamsters.** *Christina Winnicker and Brianna Gaskill*
12:30-12:45	**Can we use prepartum feeding behavior to identify dairy cows at risk for transition health disorders?** *Marcia Endres, Karen Lobeck-Luchterhand, Paula Basso Silva and Ricardo Chebel*	**Play in rats: association across contexts and types, and analysis of structure.** *Luca Melotti, Jeremy D. Bailoo, Eimear Murphy, Oliver Burman and Hanno Wurbel*
12:45-13:00	**Stability and welfare in social networks based on nearest neighbours; do group housed calves have preferential relationships?** *Paul Koene and Bert Ipema*	**Latency to start interaction after working for the resource in blue foxes.** *Tarja Koistinen, Elina Hamalainen and Jaakko Mononen*

Wednesday July 30

13:00-13:15	**Automated measurement of sheep movement order: consistent, stable and useful to identify the risk of welfare compromise?** *Amanda K Doughty and Geoff N Hinch*	**Is the more the merrier true for young mink?** *Lina Olofsson, Steffen W Hansen and Lena Lidfors*
13:15-13:30	**Moving Geographic Information Systems research indoors: spatiotemporal analysis of agricultural animals.** *Courtney L. Daigle, Debsmit Banerjee, Robert A Montgomery, Subir Biswas and Janice M Siegford*	**Investigating anhedonia in a non-conventional species: are some riding horses depressed?** *Carole Fureix, Cleo Beaulieu, Celine Rochais, Soizic Argaud, Severine Henry ,Margaret Quinton, Martine Hausberger and Georgia Mason*
13:30-14:45	Lunch	
14:45-15:30	Coffee served in Poster Session (1)	
15:30-16:15	**PLENARY: New approaches for modelling and analysis of animal movement behavior.** *Edward Codling, Jorge Vazquez Diosdado, Jonathan R Amory, Zoe E Barker, Darren P Croft and Nick J Bell*	
16:15-17:00	SESSION 3: Movement and use of space	SESSION 4: Free papers II
16:15-16:30	**The sow's udder: a neonatal piglet's foraging territory, battlefield and 'dance floor'.** *Janko Skok and Dejan Škorjanc*	**Factors influencing the inter-group agonistic behaviour of free-ranging domestic dogs (Canis familiaris).** *Sunil Kr Pal*
16:30-16:45	**Mother-infant social interactions in rangeland-raised beef cattle.** *Mohammed N. Sawalhah, Andres F. Cibils and Vanessa J. Prileson*	**Heritability of human directed social behaviours in beagles.** *Mia Persson, Lina Roth, Martin Johnsson and Per Jensen*
16:45-17:00	**Validation of HOBO Pendant® accelerometers for automated step detection in adult male turkeys.** *Hillary A. Dalton, Benjamin J. Wood, James P. Dickey and Stephanie Torrey*	**Red Jungle fowl (Gallus gallus) selected for low fear of humans are more dominant and show signs of improved welfare.** *Beatrix Agnvall, Anser Ali, Sara Olby and Per Jensen*
17:00-17:30	Coffee Break	
17:30-18:30	SESSION 3: Movement and use of space (Cont.)	SESSION 5: Cognition
17:30-17:45	**Use of a bird-mounted tri-axial accelerometer to quantify experienced physical forces during descent in laying hens.** *Janja Širovnik, Francesca Booth, John F. Tarlton and Michael J. Toscano*	**Assessment of cattle motivation for access to pasture or feedlot environments.** *Caroline Lee, Andrew Fisher, Ian Colditz, Jim Lea and Drewe Ferguson*
17:45-18:00	**Personality traits of high, low and non-users of a free range area in laying hens.** *Carlos E. Hernandez, Caroline Lee, Drewe Ferguson, Tim Dyall, Sue Belson, Jim Lea and Geoff Hinch*	**Social housing reduces fear of novelty and improves reversal learning performance in dairy calves.** *Rebecca K. Meagher, Rolnei R. Daros, Joao H.C. Costa, Marina A.G. Von Keyserlingk and Daniel M. Weary*

Wednesday July 30

18:00-18:15	**Analysis of movement in pullets: benefits of larger groups.** *Guiomar Liste, Irene Campderrich and Inma Estevez*	**Cognitive bias in gestating sows.** *Kristina Horback, Megan Murray and Thomas Parsons*
18:15-18:30	**Does LED-lighting in broiler houses improve welfare?** *Anja Brinch Riber*	**Investigating the reward cycle for play in pigs.** *Lena Lidfors, Negar Farhadi, Claes Anderson and Manja Zupan*
18:30-19:30	SESSION 6: Social networking and modeling	SESSION 5: Cognition (Cont.)
18:30-18:45	**Social dynamics of mice living in complex laboratory environments.** *Becca Franks and James P. Curley*	**Laboratory rats enjoy wheel running.** *Sylvie Cloutier, Alicia K. Netter and Ruth C. Newberry*
18:45-19:00	**Emergence of cooperation in heterogeneous populations.** *Ramon Escobedo and Annick Laruelle*	**Chronic stress leads to an enhanced stress response and an increased expectation of a food reward in sheep.** *Else Verbeek, Drewe Ferguson, Ian Colditz and Caroline Lee*
19:00-19:15	**Preferential associations between laying hens in nests.** *Nadine Ringgenberg, Ersnt K.F. Froehlich, Alexandra Harlander-Matauschek, Micheal, J. Toscano, Hanno Wurbel and Beatrice A. Roth*	**Effects of prenatal stress and post-weaning enrichment on spatial learning performances in lambs.** *Marjorie Coulon, Alain Boissy, Raymond Nowak and Frederic Levy*
19:15-19:30	**Social dominance in female beef cattle: determined by body weight or by respect for age?** *Marek Špinka, Radka Šarova, Francisco Ceacero and Radim Kotrba*	**Investigating depression-like condition in horses: attentional deficits towards auditory stimuli in some riding horses.** *Celine Rochais, Severine Henry, Carole Fureix, Cleo Beaulieu and Martine Hausberger*
19:30-21:00	Wine and Cheese Networking Event	

Thursday July 31

08:00-09:00	Editorial Board ABBS Meeting (Board members)	
09:00-10:00	**PLENARY: Ruth Harrison's Animal Machines 50 years on: still providing lessons for the future** *Marian Dawkins*	
10:00-10:45	**PLENARY: 50 years on: are animals still machines?** *Michael Mendl*	
10:45-11:45	Coffee Break and Poster Session (2)	
11:45-13:30	SESSION 7: From pain to positive emotions	SESSION 8: Methods in animal behaviour and welfare research	SESSION 9: Companion animals
11:45-12:00	**What is an emotion? The possibilities and pitfalls of reinforcement-based definitions of emotion in animals.** *Elizabeth Paul and Michael Mendl*	**Non-invasive methods based on non-linear analyses to monitor fish behavior and welfare.** *Harkaitz Eguiraun, Karmele Lopez De Ipina and Iciar Martinez*	**Application of a dog personality questionnaire to investigate personality development in juvenile guide dogs.** *Naomi Harvey, Martin Green, Gary England and Lucy Asher*

12:00-12:15	**Anxious animals? A potential method for assessing awareness of emotions in nonhuman animals.** *Walter Sanchez-Suarez, Liz Paul, Melissa Bateson and Georgia Mason*	**Infrared thermography as a measure of temperament in beef cattle: some technical constraints.** *Pol Llonch, Matt Turner, Malcolm Mitchell and Simon Turner*	**Owner perceived closeness and owner perceived intelligence of dog can predict dog's performance on object choice task.** *Jessica Lee Oliva, Jean-Loup Rault, Belinda Appleton and Alan Lill*
12:15-12:30	**Assessment of different methods to identify pain in horses.** *Diana Stucke, Sarah Hall, Beatrice Morrone, Mareile Grose Ruse and Dirk Lebelt*	**Hair cortisol as an indicator of chronic stress in growing pigs.** *Nicolau Casal, Xavier Manteca, Anna Bassols, Raquel Pena and Emma Fabrega*	**Using judgment bias to assess positive affect in dogs.** *Rebecca E. Doyle, Tegan Primmer, Rachel Casey, Rafael Freire, Sharon Nielsen and Michael Mendl*
12:30-12:45	**Lambs show ear posture changes when experiencing pain.** *Mirjam Guesgen, Ngaio Beausoleil, Edward Minot, Mairi Stewart and Kevin Stafford*	**Assessment of fear in turkeys: test reliability and correspondence.** *Marisa Erasmus and Janice Swanson*	**Behavioural and physiological indicators of affective valence and arousal in companion dogs.** *Emma Buckland, Charlotte Burn, Holger Volk and Siobhan Abeyesinghe*
12:45-13:00	**The facial expression of sheep with footrot and its relationship with lameness, total lesion score and faecal cortisol.** *Krista Mclennan, Carlos Rebelo, Mark Holmes, Murray Corke and Fernando Constantino-Casas*	**The transect method as an on-farm welfare assessment of commercial turkeys.** *Joanna Marchewka, Inma Estevez, Giuseppe Vezzoli, Valentina Ferrante and Maja M. Makagon*	**Does dog behaviour and physiology after doing wrong indicate guilt, shame or fear?** *Carla M.C. Torres-Pereira and Donald M. Broom*
13:00-13:30	SESSION 7: From pain to positive emotions (Cont.)	SESSION 10: Free paper III	SESSION 11: Welfare in wildlife
13:00-13:15	**Pain sensitivity and lying behaviour in dairy cows with different types of hock alterations.** *Christoph Winckler, Erin Mintline and Cassandra Tucker*	**Does routine human-animal-contact in beef and dairy heifers alter their reaction to a standardised veterinary examination?** *Katharina L Graunke and Jan Langbein*	**What makes grizzly bears optimistic?** *Heidi A. Keen, O. Lynne Nelson, Charles T. Robbins, Marc Evans, David J. Shepherdson and Ruth C. Newberry*

Thursday July 31

13:15-13:30	**Do ear postures indicate positive emotional state in dairy cows?** *Helen Proctor and Gemma Carder*	**Influence of castration method and sex on lamb behaviour, the development of ewe-lamb bonding and stress responses.** *Katarzyna Maslowska, Eurife Abadin and Cathy M Dwyer*	**Risk factors for poor welfare in *Psittaciformes*: do intelligence and foraging behaviour predicts vulnerable parrot species?** *Heather Mcdonald Kinkaid, Yvonne Van Zeeland, Nico Schoemaker, Michael Kinkaid and Georgia Mason*
13:30-14:45	Lunch		
13:30-15:30	ISAE Regional Secretaries Lunch		
14:45-15:30	Coffee served in Poster Session (2)		

15:30-16:15 PLENARY: Observer bias in applied ethology: can we believe what we score, if we score what we believe? *Frank Tuyttens, Eva Van Laer, Lisanne Stadig, Leonie Jacobs, Jasper Heerkens, Sophie De Graaf and Bart Ampe*

16:15-17:00	SESSION 10: Free paper III (Cont.)	SESSION 11: Welfare in wildlife (Cont.)
16:15-16:30	**Behavioral, body postural and physiological responses of sheep to simulated sea transport motions.** *Eduardo Santurtun, Valerie Moreau and Clive J.C. Phillips*	**Environmental enrichment in zoos – is the environment always enriched?** *Maria Andersson, Anna Lundberg, Claes Anderson and Madeleine Hjelm*
16:30-16:45	**Communal rearing of rabbits affects space use and behaviour upon regrouping as adults.** *Stephanie Buijs, Luc Maertens and Frank A.M. Tuyttens*	**Environmentally-enriching mink increases lymphoid organ weight and differentiates between two forms of stereotypic behavior.** *Maria Diez-Leon and Georgia Mason*
16:45-17:00	**Social behaviour of cattle in monoculture and silvopastoral systems in the tropics.** *Lucia Amendola, Francisco Solorio, Carlos Gonzalez-Rebeles, Ricardo Amendola-Massiotti and Francisco Galindo*	**Differences in neophobia between hand-reared and parent-reared barn owls (*Tyto alba*).** *Luis Lezana, Ruben Hernandez-Soto, Maria Diez-Leon, Rupert Palme and David Galicia*
17:00-17:30	Coffee Break	

17:30-19:30: WORKSHOP 1	17:30-19:30: WORKSHOP 2	17:30-19:30: WORKSHOP 3
Observer bias in applied ethology: can we believe what we score, if we score what we believe? *Frank Tuyttens, Eva Van Laer, Lisanne Stadig, Leonie Jacobs, Jasper Heerkens, Sophie De Graaf and Bart Ampe*	**Quality of life and behavioural problems in small animals.** *J. Yeates, M. Amat and T. Camps*	**Bioethics of (wildlife) research – the importance of training.** *Anna S Olsson and Tomás Redondo*

Friday August 1

09:00-09:45	**PLENARY: Enhancing cognitive function in rodents.** *Shira Knafo*
09:45-10:30	**PLENARY: The architecture of sexual behaviour, lessons from the chicken.** *Tom Pizzari*
10:30-11:30	Coffee Break and Poster Session (3)
11:30-11:45	SESSION 12: Neurobiology of behaviour and welfare

Time	SESSION 12: Neurobiology of behaviour and welfare	SESSION 13: Free papers IV
11:30-11:45	**Domestication related anxiety genes identified in chickens.** *Per Jensen, Martin Johnsson and Dominic Wright*	**Food neophobia declines with social housing in dairy calves.** *Joao H C Costa, Rolnei R Daros, Marina A G Von Keyserlingk and Daniel M Weary*
11:45-12:00	**Can peripheral nerve damage caused by tail docking lead to tail pain later in the life of pigs?** *Mette S. Herskin, Pierpaolo Di Giminiani, Dale Sandercock, Armelle Prunier, Celine Tallet, Matt Leach and Sandra Edwards*	**Operant conditioning of urination by dairy calves.** *Alison Vaughan, Anne Marie De Passille, Joseph Stookey and Jeffrey Rushen*
12:00-12:15	**Tail biting in pigs: (in)consistency, blood serotonin, and responses to novelty.** *Winanda W. Ursinus, Cornelis G. Van Reenen, Inonge Reimert, Bas Kemp and J. Elizabeth Bolhuis*	**The effect of pair housing and enhanced milk feeding on play behaviour in dairy calves.** *Margit Bak Jensen, Linda Rosager Duve and Daniel M. Weary*
12:15-12:30	**Factors affecting oxytocin concentration and relationship between oxytocin concentration and affiliative behavior.** *Siyu Chen, Shigefumi Tanaka, Sanggun Roh and Shusuke Sato*	**To brush or not to brush? The use of mechanical brushes by dairy cows and its association with agonistic events.** *Daiana de Oliveira, Therese Rehn, Yezica Norling and Linda J. Keeling*
12:30-12:45	**Physiological difference between crib-biters and control horses in a standardised ACTH challenge test.** *Sabrina Briefer Freymond, Deborah Bardou, Elodie F. Briefer, Rupert Bruckmaier, Natalie Fouche, Julia Fleury, Anne-Laure Maigrot, Alessandra Ramseyer, Klaus Zuberbuehler and Iris Bachmann*	**Social status and tear staining in nursery pigs.** *Amelia E.M. Marchant-Forde and Jeremy N. Marchant-Forde*
12:45-13:00	**Vocalization latency of neonate lambs as an indicator of vigour.** *Christine Morton, Geoffrey Hinch, Drewe Ferguson and Alison Small*	**Analysis of the phenotypic link between aggression at mixing and long-term social stability in groups of pigs.** *Suzanne Desire, Simon P. Turner, Richard B. D'Eath, Andrea Doeschl-Wilson, Craig R.G. Lewis and Rainer Roehe*

Friday August 1

13:00-13:15	**How sheep in different mood states react to social stimuli varying in valence.** *Sabine Vogeli, Beat Wechsler and Lorenz Gygax*	**Horse welfare: the use of Equine Appeasing Pheromone (EAP) during clipping to facilitate the adaptation process.** *Alessandro Cozzi, Elisa Codecasa, Olivier Carette, Jerome Arnauld Des Lions, Thomas Normand, Florence Articlaux, Celine Lafont-Lecuelle, Cecile Bienboire-Frosini and Patrick Pageat*
13:15-15:15	Lunch	
14:30-15:15	Coffee served in Poster Session (3)	
15:15-17:15	SESSION 14: Behaviour and reproduction	SESSION 15: Free papers V
15:15-15:30	**Maternal investment and piglet survival – behavioural and other maternal traits important for piglet survival.** *Inger Lise Andersen*	**Olfaction – the forgotten sense (in applied ethology).** *Birte L Nielsen*
15:30-15:45	**Farrowing progress and cortisol level in sows housed in SWAP pens.** *J Hales, VA Moustsen, MBF Nielsen, PM Weber and CF Hansen*	**Attitudes amongst veterinary students towards animal sentience: cross sectional and longitudinal findings.** *Nancy Clarke, Liz Paul and David Main*
15:45-16:00	**Calf weaning in a partial suckling system.** *Julie Foske Johnsen, Annabelle Beaver, Cecilie Mejdell, Anne Marie De Passille, Jeffrey Rushen and Dan Weary*	**Effect of aviary housing characteristics on laying hen welfare and performance.** *Jasper L.T. Heerkens, Ine Kempen, Johan Zoons, Evelyne Delezie, T. Bas Rodenburg, Bart Ampe and Frank A.M. Tuyttens*
16:00-16:15	**Neonatal lamb rectal temperature, but not behaviour at handling, predict lamb survival.** *Cathy M Dwyer, Susanne Lesage and Susan E Richmond*	**Feather pecking in laying hens during rearing and laying in non-cage systems, management practices and farmers opinions.** *Elizabeth Nicole (Elske) De Haas and Bas T Rodenburg*
16:15-16:30	**Prenatal stocking densities affect fear responses and sociality in goat (*Capra hircus*) kids.** *Rachel Chojnacki, Judit Vas and Inger Lise Andersen*	**Solitary play in offspring of laying hens with high and low feather-pecking activity.** *Stephanie Bourgon, Margaret Quinton and Alexandra Harlander-Matauschek*
16:30-16:45	**Pregnancy enhances the frequency of vocalizations in rabbit does expose to buck.** *Imene Iles*	**Are keel bone fractures in laying hens related to bone strength or to fearfulness?** *T. Bas Rodenburg, Jasper L.T. Heerkens and Frank A. M. Tuyttens*
16:45-17:00	**On-farm evaluation of the copulation act in camels (*Camelus dromedaries*).** *Mohamed M.A. Mohsen, S.A. Al-Shami, Marzouk A. Al-Ekna, Saad A. Al-Sultan and Ahmad A.A Alnaeen*	**Managing non-beak trimmed hens in furnished cages: using two strains of laying hens and extra environmental enrichment.** *Krysta Morrissey, Tina Widowski, Laurence Baker and Victoria Sandilands*

Friday August 1

17:00-17:15	**Hatch time affects behavior and weight development in White Leghorn laying chickens.** *Pia Lotvedt and Per Jensen*	**Changing the animal or the environment: changes in breeding strategy and housing conditions to improve the welfare of pigs.** *J. Elizabeth Bolhuis, Inonge Reimert, Winanda W. Ursinus, Irene Camerlink, T. Bas Rodenburg and Bas Kemp*
17:15-17:45	Closing Ceremony	
17:45-19:15	ISAE General Meeting	
21:00-01:00	**Congress Banquet** *´El Caserón´* **– Armentia**	

Saturday August 2

EXCURSIONS (Full day, see departure time for each excursion)

Table of contents

Session 1. Precision farming

Session 2. Free papers I

Session 3. Movement and use of space

Session 4. Free papers II

Session 5. Cognition

Session 6. Social networking and modeling

Session 7. From pain to positive emotions

Session 8. Methods in animal behaviour and welfare research

Session 9. Companion animals

Session 10. Free paper III

Session 11. Welfare in wildlife

Session 12. Neurobiology of behaviour and welfare

Session 13. Free papers IV

Session 14. Behaviour and reproduction

Session 15. Free papers V

Poster session 1

Poster session 2

Poster session 3

Wood-Gush Memorial Lecture: Wednesday July 30, 8:30-10:00

Agent-based modelling: a powerful tool for applied ethology

Volker Grimm
Helmholtz Centre for Environmental Research, UFZ, Permoserstr. 15, 04318 Leipzig, Germany; volker.grimm@ufz.de

Agent-based models (ABMs) explicitly represent decision-making agents and their interactions with each other and their environment. ABMs are used to explore the mutual relationship between the agents' behaviour and the properties of the system they are part of. Examples include cells in tissues, organisms in ecological systems, or brokers in financial markets. ABMs go beyond more aggregated mathematical models as they allow to simultaneously represent variation among individuals, local interactions, and adaptive behaviour. ABMs are implemented as computer programs. Specific software platforms exist which facilitate implementation and testing. A standard format for verbally describing ABMs – the ODD protocol – got established in ecology over the last five years and started making the design of ABMs more coherent and efficient. ABM description should include the model's purpose, its entities and state variables, its processes and how they are scheduled, the concepts underlying the model's design, and details about input data, initialisation, and the submodels representing the processes. To make sure ABMs capture the internal organisation of their real counterparts sufficiently well, they are designed to simultaneously reproduce multiple patterns observed in the real system at different hierarchical levels and scales. This strategy of pattern-oriented modelling is used to optimize model complexity, to inversely determine sets of unknown parameters, and to select among alternative submodels representing specific processes. In contrast to behavioural sciences, ABMs do not focus on behaviour per se, but try to find representations of decision making and behaviour that are good enough to make the ABM reproduce system-level patterns. In ecology, first principles driving behaviour are energy budgets and fitness seeking. Recently, also emotions and different personalities were included. Once an ABM has been developed and tested, it can be used as a virtual laboratory to perform simulation experiments. This holds great potential also for applied ethology, as ABMs allow exploring scenarios which cannot be implemented in reality because of ethical, logistic, or financial constraints. For example, a major task would be to model sensing: over what distances and in what detail does an animal perceive its biotic and abiotic environment? Once such a model is established, it can be used to explore how various factors such as group size and composition, environmental parameters, and features of the artificial habitat affect the animals welfare. I will present examples from various disciplines and outline the potential use of ABMs in applied ethology and discuss the kind of data needed for developing and testing such ABMs.

Precision livestock farming to improve animal welfare and health

Daniel Berckmans

Division M3-BIORES, Department of Biosystems, Kasteelpark Arenberg 30, 3001 Heverlee, Belgium; daniel.berckmans@biw.kuleuven.be

The worldwide demand for meat and animal products is increasing while the number of farmers is decreasing. This results in bigger farms with more animals to be managed by a single farmer. A main problem is that the farmer has not enough time to follow and monitor each individual animal. Meanwhile farmers are stressed due to a lack of control and follow up of their animals. The main objective of Precision Livestock Farming (PLF) is to deliver new management systems based upon continuous fully automated monitoring of animals by using modern ICT technology (sensors, image analysis, and sound analysis). By monitoring animal behaviour (eating and drinking behaviour, animal activity, spatial distribution, aggressive behaviour, playing behaviour, weight, water and feed intake, etc.) the animal welfare, health and production parameters can be followed. The systems based on cameras can easily analyse 5 images per second while sound monitoring systems can analyse up to 40,000 measured sound samples per second. This means that animals are really monitored on a continuous basis that is more consistent that what a farmer ever was able to do. Other advantages of PLF-technology are that the system is continuously operational all day and night; since less farm visits by experts are necessary (e.g. Welfare Quality approach), there is reduced risk for disease transfer, the observation system is part of the farmers' management system and should be an investment with interesting payback period; the farmers gets early warnings and can save time with more quietness regarding the continuous follow up of this animals. This paper shows examples of systems that have already been developed in order to explore and demonstrate the potential of this technology. Examples are be given on cows, pigs and broilers where fully automated monitoring and real time data analyses give early warnings and scores on condition and health of animals. For milking cows a fully automated lameness monitor has been developed. Each time that the cow is going to the milking robot she is walking under a camera that allows calculating the individual gait analysis for this cow. For fattening pigs a sound analysis systems has been developed to detect starting coughing of animals. For broilers a camera system has been developed that measures continuously the activity of the birds and their distribution in the broiler house of 60,000 birds. The system detects over 90% of management problems that occur and gives a warning to the farmer. It is clear that the technology is progressing and this always to focus more on detailed analysis of animal behaviour to measure and interpret their welfare, health and production status.

New approaches for modelling and analysis of animal movement behaviour

Edward Codling[1], Jorge Vázquez Diosdado[1], Jonathan R Amory[2], Zoe E Barker[2], Darren P Croft[3] and Nick J Bell[4]

[1]*University of Essex, Department of Mathematical Sciences, Colchester, Colchester, United Kingdom,* [2]*Writtle College, Centre for Equine and Animal Sciences, Chelmsford, Chelmsford, United Kingdom,* [3]*University of Exeter, Department of Psychology, Exeter, Exeter, United Kingdom,* [4]*Royal Veterinary College, Dorchester, Dorchester, United Kingdom; ecodling@essex.ac.uk*

Over the past two decades there has been a technological revolution in the availability of reliable trackers, loggers and sensors for recording animal movement and behaviour at high temporal resolution for long periods of time. This has subsequently led to the collection of enormously rich movement data sets that have required the development of new tools and methods of analysis. Understanding animal movement is an important biological question in its own right, but the wealth of data now available may also allow us to answer more general questions about animal behaviour across different temporal and spatial scales. In this plenary talk we will review the traditional methods used for modelling and analysing individual animal movement paths, covering topics such as random walks, correlated and oriented movement, and movements within a confined space or home range. We will discuss new methods and approaches for the analysis of movement data that allow us to determine different types of animal space use and behavioural state. These include direct analysis of movement statistics through control charts, space use analysis using Brownian bridges, and detection of behavioural changes using hidden Markov models. We will also present a range of potential approaches for analysing dynamic social interactions in an animal group. The talk will conclude with a discussion of how the analysis of individual- and group-level animal movement behaviour could be linked together under a common framework. The various modelling and analysis approaches will be illustrated using an example data set obtained by tracking dairy cow movements within a commercial barn. We will demonstrate how analysis of cow movement behaviour and social interactions could lead to early identification of health and welfare issues within the herd.

Ruth Harrison's Animal Machines 50 years on: still providing lessons for the future

Marian Dawkins
University of Oxford, Zoology, South Parks Road, Oxford, United Kingdom;
marian.dawkins@zoo.ox.ac.uk

The publication of Ruth Harrison's book Animal Machines in 1964 was a landmark in the history of animal welfare. The public was alerted to new farming methods, such as keeping hens in battery cages that had been developed to improve food production. The book was widely quoted in the popular press and raised the debate about farm animal welfare to new levels. Many people believe that current legislation giving protection to food animals introduced in the EU and around the world can be traced back to this pioneering book. But Animal Machines is of much more than just historical interest. What is striking about it is the many different arguments that Ruth Harrison. She described the risks to human health of keeping animals in confined dirty conditions and the problems of waste disposal caused to the environment. There is a whole chapter about money and the financial advantages of improving animal welfare. She did not hesitate to use every argument she could lay her hands on t and in this we still have much to learn from her book. We are now entering a new and dangerous era in which farming is under pressure to become even more 'efficient' and 'sustainably intensive'. We need to make sure that this is not at the expense of animal welfare. I shall argue that we can best do this by following the lead that Animal Machines gave us 50 ago. Using a variety of examples from current animal welfare science I shall show how 'efficiency', technology and high welfare can coexist but that this will only happen if we develop solutions that are good for the environment, good for human health and work commercially so that farmers can make a better living by improving animal welfare than by ignoring it.

50 years on: are animals still machines?

Michael Mendl

University of Bristol, School of Veterinary Science, Langford House, Langford, BS40 5DU, United Kingdom; mike.mendl@bris.ac.uk

Ruth Harrison's ground-breaking 1964 book, Animal Machines, was the catalyst for the emergence of animal welfare science. It provoked the British Government to set up the Brambell Committee which went on to publish the Brambell Report including an insightful appendix by ethologist W.H. Thorpe raising many prescient questions about the assessment of pain and distress in animals, and calling for new research on animal welfare. Fifty years on, are animals still machines? In one sense, yes. Many of the husbandry systems criticised by Ruth Harrison remain in place in various parts of the world – the treatment of farm animals as if they are little more than food producing mechanisms continues. The idea that non-human animals actually are no more than Cartesian machines – without a mind – also persists amongst some people and for some/many species and is one driver of the tolerance of highly intensive livestock production. Nevertheless, the scientific study of animal mind has blossomed and the information about animal cognition and emotion that it generates has the potential to influence attitudes towards 'animal machines' and hence to drive changes in legislation and the public's response to animal welfare issues. If probing the animal mind remains a valuable endeavour in animal welfare science, what are the challenges and opportunities? Cataloguing complex cognitive abilities is common in animal cognition research and examples of human-like behaviour captivate the public. However, what they actually mean from an animal welfare perspective requires critical consideration. Do clever animals suffer more? Do particular cognitive abilities help or hinder responses to welfare challenges, and can they be used to manage animals more humanely? As Ruth Harrison herself noted, the ability to experience emotion or pain is likely to be especially important for an animal's welfare, and studies aimed at understanding and measuring animal emotion are proliferating. Again we need to carefully question what this research shows and be wary of sliding into anthropomorphism. Clear operational definitions of emotion are needed, alongside the realisation that understanding the conscious experiences of other species remains a major unsolved challenge. Defining emotion in terms of underlying reward and punishment systems may be more practically and theoretically illuminating than thinking of them as human-like discrete emotions. Studies of metacognition, mental time travel and recall memory, and emerging techniques such as brain imaging and optogenetics may shed new light on mental processes in other species. Even if we cannot solve the consciousness question, carefully targeted new information about animal minds can be of practical, theoretical and educational value in the animal welfare debate.

Observer bias in applied ethology: can we believe what we score, if we score what we believe?

Frank Tuyttens[1,2], Eva Van Laer[2], Lisanne Stadig[2], Leonie Jacobs[2], Jasper Heerkens[2], Sophie De Graaf[2] and Bart Ampe[2]
[1]Ghent University, Faculty of Veterinary Medicine, Salisburylaan 133, 9820 Merelbeke, Belgium, [2]Institute for Agricultural and Fisheries Research (ILVO), Animal Sciences Unit, Scheldeweg 68, 9090 Melle, Belgium; frank.tuyttens@ilvo.vlaanderen.be

Rating animal behaviour and condition is the methodological cornerstone of applied ethology. As blinding observers is not common and often not feasible, the data collection staff is usually aware of relevant information about the animals they observe. This information may evoke expectations about the rating outcome which, in turn, may lead to conscious or unconscious biases in the data, i.e. expectation bias. In order to start documenting the extent to which this type of observer bias may pose a problem in applied ethology, we investigated whether observer expectations influenced subjective scoring methods during a class practicum. Veterinary students (n=157) were briefly trained in (1) assessing laying hen behaviour using Qualitative Behaviour Assessment (QBA), (2) scoring the degree of panting to assess heat stress in cattle, and (3) recording negative and positive interactions between pigs. The students were divided into two groups and applied these methods in three experiments. During these experiments they were shown duplicated video recordings of the same animals: the original and a slightly modified version (to prevent recognition at second viewing). Prior to scoring they were either told correct or false information about the conditions in which the animals had been filmed. One group was given false information for the modified recording and correct information for the original recording, and vice versa for the other group. The false information reflected plausible study scenarios in ethology and aimed to create expectations about the outcome. As in reality they scored the identical behaviour twice, the difference in scores when given true versus false information reflects expectation bias. In all experiments there was evidence of expectation bias: the QBA indicated more positive and fewer negative emotions when told that the laying hens were from an organic instead of a conventional farm, the cattle panting score was higher when told that the ambient temperature was 5 °C higher than in reality, and the ratio of positive to negative interactions was higher when told that the observed pigs had been selected for high social breeding value. Although veterinary students may not be representative of practicing ethologists, these findings do indicate that observer bias could influence subjective scores of animal behaviour and welfare. We discuss the implications of these findings in the field of applied ethology and suggest possible solutions.

Enhancing cognitive function in rodents
Shira Knafo
UPV/EHU, Unidad de Biofísica, Barrio Sarriena s/n, 48940 Leioa, Spain; s.knafo@ikerbasque.org

We are studying the molecular mechanisms of processes of learning and memory in rodents in normal conditions and in conditions of cognitive impairment and enhancement. Behavioural research suggests that changes in cognitive function can be reliable indicators of animal welfare. Therefore, it is important to improve cognition in animals in any setting. On the first part of the talk I will describe how housing laboratory animals in stressful conditions (such as chronic individual caging or overcrowding) at any phase of their lives results with changes in brain structures involved in cognition and mental health. Additionally, research now relays exposure of laboratory animals to stress in early life with increased sensitivity to stress and to cognitive deficits in adulthood. This effect on cognitive function probably arises from remodelling of brain architecture and loss of neurons in the hippocampus, the brain area essential to learning and memory and from dendritic atrophy and loss synaptic contacts. The second part of the talk will be dedicated to the description of approaches for cognitive enhancement. In 'enriched' environment the animals are kept in large cages and in bigger groups enabling more natural social interaction. The environment contains toys, tunnels, nesting material, food and running wheels and is altered periodically. Specifically I will describe how when the environment is 'enriched' in relation to standard laboratory housing conditions, rodents present better cognitive function, probably due to larger hippocampus that contains more neurons. Importantly environmental enrichment can also revert negative consequences of early life stress in rodents. I will also describe the cellular correlates of cognitive enhancement in rodents. Thus, the cognitive function of laboratory animals is largely controlled by their housing conditions, a factor that eventually determines their welfare.

The architecture of sexual behaviour, lessons from the chicken

Tom Pizzari
Edward Grey Institute, Department of Zoology, University of Oxford, OX1 3PS Oxford, United Kingdom; tommaso.pizzari@zoo.ox.ac.uk

Fertility and welfare are key aspects of modern animal production programmes, yet they are often considered from a largely proximate perspective and in isolation from each other. For example, fertility has often been considered an entirely male or female physiological trait, the additive genetic basis of which has been the targets of considerable artificial selection efforts. However, this approach fails to acknowledge a large source of variance in fertility due to behavioural mechanisms, which modulate male and female sperm allocation and utilisation. The overarching goal of this contribution is to illustrate how evolutionary theory allows a more holistic functional approach in which reproductive performance and welfare are inter-dependent. I review recent experimental studies conducted on populations of red junglefowl, the wild ancestor of the domestic chicken, as well as different domestic chicken strains to show how reproductive performance results from the complex interactions of sophisticated strategies of sexual behaviour moulded by a number of evolutionary processes: sexual selection, sexual conflict and kin selection. In particular, reproductive behaviours are often best understood as evolutionary adaptations to the selective pressures imposed by sperm competition, a scenario in which the ejaculates of different males compete to fertilise the same set of eggs, a mating system prevalent in the ancestral red junglefowl, and often favoured by domestication. These strategies also have important implications for the welfare of individual birds and the productivity of the population as a whole. I then illustrate some examples of how in the near future this heuristic approach can be generalized and applied to develop practices, which simultaneously maximize the productivity and optimize the welfare of managed animal populations.

Prenatal and early postnatal effects on offspring development

Inger Lise Andersen
Norwegian University of Life Sciences, Animal and Aquacultural Sciences, P.O. Box 5003, 1432 Ås, Norway; inger-lise.andersen@nmbu.no

Prenatal stressors related to farming conditions range from human handling, isolation and transport stress to social stress. Over the last 15 years, increasing studies have investigated effects of prenatal stress in farm animals, covering a wide range of species including foxes, poultry, farmed fish, small ruminants and pigs. While previously, scientists mainly focused on negative effects of prenatal stress, there are now quite a few examples indicating that mothers subjected to moderate levels of stress produce offspring well adapted to the present production environment, in accordance with the environment-adaptation hypothesis. The effects of prenatal stress depend not only on the type of stressors but also on the magnitude of the stress and its timing and duration during pregnancy. Few farm animal studies have reported effects on reproductive success, either because farm animals are generally robust or rarely abort once the foetus is implanted, or because most studies have not focused on the first days of pregnancy when abortion is more likely to occur. The most common experimental tests used to study behavioural development in the offspring are open-field, maze, separation, social and handling tests, in which, in most cases, exploratory activity and fear responses such as escape attempts, freezing and vocalisations are measured. The majority of research demonstrates that subjecting the mother to prenatal stress during the third trimester of pregnancy, results in offspring showing a more proactive behavioural style. Results on learning abilities and problem solving strategies are dominated by studies in rodents and primates. These studies suggest that, even if prenatal stress impairs learning ability when the magnitude of the stress is high, learning in fear situations may actually also be facilitated by the stress. Social strategies and flexibility in social situations as well as in problem solving have not been much in focus in farm animals. If offspring become more proactive, competitive and/or learning abilities are impaired, this could have consequences for breeding animals in more than one generation. Early postnatal experience has strong effects on offspring development and is sometimes more important than the prenatal conditions. The mother may buffer negative effects of the prenatal environment in the early postnatal period, reducing or even reversing some predicted negative effects of prenatal stress. The interaction between effects of the prenatal vs. early postnatal environment should thus be a focus of future farm animal studies. Research to date underlines the importance of understanding how the mother is coping with her environment and shaping offspring behaviour.

Satellite Workshop 2: Tuesday July 29, 16:00-18:00

Dam rearing in dairy production

Katharina A Zipp[1], Julie J Johnsen[2], Tasja Kälber[3], Cecilie Mejdell[2], Kerstin Bar[3] and Ute Knierim[1]
[1]University of Kassel, Farm Animal Behaviour and Husbandry Section, Nordbahnhofstr. 1a, 37213 Witzenhausen, Germany, [2]Norwegian Veterinary Institute, Department of Health Surveillance, P.O. Box 750 Sentrum, 0106 Oslo, Norway, [3]Thünen Institute für Ókologischen Landbau, Thünen Institute of Organic Farming, Trenthorst 32, 23847 Westerau, Germany; zipp@uni-kassel.de

In dairy production it is common to separate calves from the dams shortly after birth. There is a growing interest in systems allowing dam rearing, combining regular milking schedules and suckling. Nursed calves grow well, don't show cross-sucking and have a better social ability even as heifers. However, reduced milk ejection, barn design and stress at weaning keep many farmers from practicing dam rearing. The workshop will discuss perspectives of ongoing research and future possibilities for research in the field of dam rearing. Major topics will be milking characteristics of cows, development and behaviour of calves during contact and at weaning as well as long-term influences of dam rearing on female offspring and epidemiological production results. Aims are to summarize the state of science in this field, point up major challenges, develop questions and eventually solutions and get connected.

Quality of life and behavioural problems in small animals

J. Yeates[1], M. Amat[2] and T. Camps[2]

[1]RSPCA, Wilberforce Way, Southwater, United Kingdom, [2]Animal Nutrition and Welfare Service, Autonomous University of Barcelona (UAB), Barcelona, Spain; james.yeates@rspca.org.uk

In this seminar we will discuss the concept of quality of life (QoL) in veterinary medicine and its relationship with behavioural problems, with a particular emphasis on medical problems and lack of environmental predictability. The concept of QoL is widely used in human medicine in order to describe how people feel about their life. In general terms, QoL represents the balance between positive and negative experiences. Although the study of QoL in animals, especially in small animals, is much more recent, it is widely accepted that behavioural problems can decrease QoL in small animals. The first aim of this workshop is to discuss the concepts of QoL and its relation with the most common behavioural problems in small animals (aggressive behaviours, separation-related disorders, fears and phobias, repetitive behaviours, etc.). There are many medical conditions that decrease QoL of small animals. Many of these problems can be both a cause and a consequence of behavioural problems. For example, chronic pain can lead to behavioural problems (especially aggressive behaviours and fears and phobias). The second aim of this workshop is to assess the bidirectional relationship between medical conditions and behavioural problems and how this relationship can decrease QoL in companion animals. Finally, it is well known that lack of environmental predictability is an important factor that may lead to stress. The third objective of this workshop is to discuss the effect of environmental unpredictability in the development of certain behavioural problems in small animals and its implications for treatment.

Workshop 3: Thursday July 31, 17:30-19:30

Bioethics of (wildlife) research – the importance of training

I Anna S Olsson

IBMC, Laboratory Animal Science, Rua Campo Alegre 823, 4150-180 Porto, Portugal; olsson@ibmc.up.pt

The use of animals for research purposes has a long history of ethical controversy. In response to public concern for how animals are used in research, this activity is also tightly regulated. The main mechanisms for regulation are legislation, which lays down general rules for how animals are to be housed and handled, and the evaluation processes (e g ethics review) which deliberate on individual research projects involving animals. These mainly externally imposed mechanisms are important for setting standards and improving public trust in research. However, to a large extent, it is the attitude, knowledge and skills of individual researchers which ultimately decide how much animal welfare is taken into account when studies are planned and performed. For conventional laboratory-based animal research, there is a long tradition of developing training courses, with the first international recommendations appearing in 1995. Surveys of more than 200 course participants have shown that such training increases both knowledge and implementation of the 3Rs principle in research. However, this will of course only be efficient if the knowledge researchers acquire during training is relevant for their research. At present, regular training in wildlife research methods is not available in Europe. In this presentation, which is intended to be integrated into the workshop on bioethics in wildlife research, I will explore what such training could look like, drawing on my experience as organizer of internationally accredited training in laboratory animal science.

Cardiac activity, behavior and cortisol release in dairy cattle during interaction with scraper robots

Renate Luise Doerfler[1], Jana Ebert[1], Anne Gramsch[1], Katharina Haas[1], Sylvia Kuenz[1], Martin Pardeller[1], Heike Kliem[2], Martin Bachmaier[1] and Heinz Bernhardt[1]
[1]Technische Universität München, Center of Life and Food Sciences, Chair for Agricultural Systems Engineering, Am Staudengarten 2, 85354 Freising-Weihenstephan, Germany, [2]Technische Universität München, Center of Life and Food Sciences, Chair for Physiology, Weihenstephaner Berg 3, 85354 Freising, Germany; renate.doerfler@wzw.tum.de

Scraper robots are increasingly used on dairy farms for cleaning slatted floors. The complexity of the drive paths of the robot may reduce predictability and thus controllability of situations. Lacking control of circumstances is known to arouse stress responses in animals. It has not yet been analyzed how autonomous mobile systems affect dairy cattle. The objective of the experiment, therefore, is to analyze possible stress responses in cows when interacting with scraper robots, the process of familiarization and implications for animal welfare. The experiment was carried out with a total of 36 dairy cows. 12 focus cows were selected on three dairy farms with cubicle housing systems. Over a period of five weeks, cardiac activity (heart rate variability, heart rate) and behavior were recorded. Moreover, faecal samples for analysis of cortisol metabolites were collected. Baseline values of all parameters were compared with test values characterizing the time when the robot is used in the barn. A statistical analysis was conducted with linear mixed effects models. Initial results indicate that behavioral changes occur both in the beginning and at the end of test measurements. The duration of lying in the cubicle (P=0.07), standing in the cubicle (P=0.047) and standing in the concentrate feeder (P=0.07) differed from baseline values. Regarding cardiac activity, statistical differences appear neither in the heart rate variability variable root mean square of successive beat-to-beat differences (P=0.70) nor in heart rate (P=0.57). The retransformed means of heart rate are 86.0 (baseline). 85.8 (test period 1), 85.2 (test period 2), and 86.7 (test period 3) beats/min. Faecal cortisol metabolite concentrations did not change during the experimental period (P=0.36). Small differences in behavior provide some evidence for stress reactions in animals. Behavioral changes indicate that familiarization to the robot was not completed in the 4 week period. The results imply that dairy cows mainly have the capacity to cope with the robot and to control their environment. However, the statistical outcome my be influenced by effects of individual animals and farms and should therefore be interpreted cautiously. It is concluded that stress responses in dairy cattle in the presence of the robot and impairment of animal welfare is on a low level. Data analysis will be continued.

The animal-machine interface: dairy calves' adaptation to automated feeders

Mayumi Fujiwara[1], Jeffrey Rushen[2] and Anne Marie De Passille[2]
[1]University of Edinburgh, Edinburgh, Scotland, United Kingdom, [2]University of British Columbia, Agassiz, BC V0M 1A0, Canada; amdepassille@gmail.com

Feeding dairy calves with automated feeders has likely welfare advantages but calves' difficulties adapting to feeders may reduce these advantages. We examined factors influencing calves' adaptation to automated milk feeders. In Exp. 1, 77 calves were introduced to group pens with automated feeders 6 d after birth. Latency to first voluntary milk ingestion, milk intakes and weight gains were recorded. A large variation was found in the latency to the first voluntary milk ingestion and this was negatively correlated with 14 d milk intake (r=-0.29, P=0.02). In Exp. 2, 55 calves were housed for 8-14 d after birth either individually or as pairs in double pens and then transferred to group pens. There were no differences between the types of early housing on the latency to first voluntary milk intake after group housing (Wilcoxon test: P>0.10). The latency to first milk ingestion was negatively correlated with milk intake over 6 d in the group pens (r=-0.38, P=0.03). When data from the two experiments were combined, latency to the first milk intake was negatively correlated with age at time of introduction (r=-0.23, P=0.009) and duration of standing (measured by accelerometers) in the days after birth (r=-0.21, P=0.02). There are large differences between calves in how quickly they learn to use automated milk feeders and slow adapting calves can have reduced milk intakes. Pair housing immediately after birth did not help. Younger calves at the time of introduction are more likely to take longer to adapt but many 6 d old calves adapt quickly, especially those that show high vigour (i.e. can stand for longer) in the first week after calving. To profit from potential advantages of automation, we need to understand the factors that help animals learn to use these machines.

The accuracy of the automatic recording of lying behaviour is affected by whether dairy cows are inside or outside

S. Mark Rutter, Norton E. Atkins and Stephanie A. Birch
Harper Adams University, The National Centre for Precision Farming, Newport, Shropshire, TF10 8NB, United Kingdom; smrutter@harper-adams.ac.uk

The IceTag (IceRobotics Ltd, Edinburgh, Scotland, UK) is a popular device used by scientists for the automatic recording of lying, standing and walking behaviour. Although studies have shown good agreement between data recorded using IceTags and manual observations in housed cattle, there have not been any studies directly comparing the agreement in cows when indoors compared with on pasture. Sixteen lactating Holstein-Friesian cows were fitted with IceTags. Eight cows were taken out to pasture between morning and afternoon milking (mean times 05:11-15:15 h), and the other eight were out between afternoon and morning milking (15:15-05:11 h). Cows were in a cubicle house when not at pasture. Cows were manually observed for two four-hour periods (06:00-10:00 h and 18:00-22:00 h) and lying behaviour recorded using instantaneous sampling at five minute intervals. The total time spend lying and number of bouts of lying recorded by the IceTags was summed for each five minute period, and these were compared with the time spent lying and number of bouts from the manually observed data. The number of discrepancies between these two sets of results was analysed using an ANOVA. The discrepancies in the number of minutes of lying in each four hour period were small, and there was not a significant difference between data recorded at pasture and indoors (3.75 minutes vs 0.95 minutes, P=0.553). In contrast, there was a considerable discrepancy in the number of lying bouts recorded in each four hour period at pasture but not when indoors, and this difference in the discrepancy between pasture and indoors was statistically significant (27.2 bouts vs 1.2 bouts, P<0.001). The close agreement in the time spent lying between manual observation and the IceTags, both indoors and at pasture supports the findings of previous studies. Compared with manual observation, the IceTags made a small overestimate (1.2) in the number of bouts recorded indoors but they made a gross overestimate (27.2) in the number of lying bouts in cows at pasture. The majority of the additional lying bouts recorded by the IceTag were typically short bouts of a few seconds duration occurring when manual observation recorded the cow as standing. Although one reason for this discrepancy could be related to the use of a five minute manual observation sample interval, this discrepancy requires further research. In conclusion, whilst IceTags show close agreement with manual observation of lying times indoors and at pasture, they appear to overestimate the number of lying bouts at pasture, and caution is advised in interpreting such data until further research has resolved the issue.

Can we use prepartum feeding behavior to identify dairy cows at risk for transition health disorders?

Marcia Endres, Karen Lobeck-Luchterhand, Paula Basso Silva and Ricardo Chebel
University of Minnesota, 1364 Eckles Avenue, St. Paul, MN 55108, USA; miendres@umn.edu

The objective of this study was to investigate whether changes in prepartum feeding behavior could be used as an indicator of health disorders in transition cows. Retrospective daily feeding times for 925 prepartum Jersey dairy cows within 21 d prepartum were used. Data were from 2 studies: study 1, 209 prepartum cows enrolled in either a stable group of 44 cows moved to a pen with no new cows added during the 5-wk rep, or conventional group with cows added once weekly to maintain a desired pen stocking density of 44 cows; or study 2, 716 prepartum cows housed at either 80% (38 cow/48 headlocks) or 100% (48 cows/48 headlocks) feedbunk stocking density with twice weekly entrance of new animals. Prepartum feeding behavior (daily feeding time) was measured using 10-min video scan sampling for 24-hr periods (4 d/wk for study 1 and 2 d/wk for study 2). Blood samples were taken on days in milk (DIM) 3, 10, 17 and 24 for measurement of β-hydroxy butyrate (BHB) concentrations. Cows were classified as having subclinical ketosis when BHB levels were ≥1,400 µmol/l. All cows were examined on d 1, 4, 7, 10, and 13 for metritis and retained fetal membrane. Lameness was evaluated on DIM -28, 0, and 35; cows with locomotion score ≥3 (1-5 scale; 1=normal, 5=severely lame) were considered lame. Cows with a health disorder were excluded from the lameness analysis. Other health events were obtained from farm records. The MIXED procedure of SAS was used to determine whether daily feeding time was associated with transition health disorders. Study was not signifcant in the model, therefore data from thc 2 studies were combined for analysis. There was a reduction in daily prepartum feeding time for cows with metritis (difference, LSMEANS ± SE min/d; 7.4±2.7; P<0.01), subclinical ketosis (15.4±7.2; P=0.03), retained fetal membrane (8.8±3.9; P=0.02), mastitis (9.2±4.3; P=0.03), lameness at DIM 0 (56.2±10.0; P<0.01) and lameness at DIM 35 (25.0±6.2; P<0.01) compared to cows that did not have the corresponding disorder. There was a tendency for a reduction in feeding time for cows with displaced abomasum (P=0.09) and twin pregnancy (P=0.08). In conclusion, prepartum feeding behavior appears to be a useful indicator of cows at risk for transition disorders. Automated feeding behavior monitoring sensors are commercially available in the USA today which will allow dairy producers to monitor transition cows to improve their welfare.

Stability and welfare in social networks based on nearest neighbours; do group-housed calves have preferential relationships?

Paul Koene and Bert Ipema
Wageningen Livestock Research, Animal Welfare, De Elst 1, 6708 WD Wageningen, the Netherlands; paul.koene@wur.nl

Social networks in animal species may indicate social stress and social support that are important for fitness, survival and welfare. Data on the social networks of animal species in captivity are mostly lacking and are needed to recognize the importance of a network for individual welfare. Social Network Analysis (SNA) allows to characterize groups, subgroups, transfer of information and individual actions and preferences. We describe small social networks based on nearest neighbours in horses, bears, chicken and veal calves to illustrate the increasing importance of measuring their social network for animal management. Validation of a network is shown in a group of mares and foals based on nearest neighbours with behavioural observations of social interactions (Nearest neighbour matrix correlates with the allogrooming matrix, $P<0.001$). SNA information is used to guide welfare-friendly removal of foals from the herd. Pitfalls and opportunities are investigated further in the network of a group of brown bears using information of solitary and territorial individuals. The social network of chicken is investigated using networks related to conspecifics (1-mode) and the available facilities for eating, drinking, scratching and perching (2-mode). SNA helped in identifying feather pecking relations. Finally, we show and emphasize the automatic measurement of location and/or nearest neighbours for management purposes using data on veal calves. 'Calves seem to form preferential relationships before 3.5 months of age. Keeping cattle together from an early age seems beneficial for them'. However, SNA in veal calves in a pen with cubicles, lying area and feeding places showed significant daily changes in their social network, indicating instability of the social relations (correlation between daily networks are not significant ($P>0.05$). We conclude that social networks are important for the welfare management of captive animal species and that automatic recording and follow-up management actions are feasible in the near future.

Automated measurement of sheep movement order: consistent, stable and useful to identify the risk of welfare compromise?

Amanda K Doughty and Geoff N Hinch
University of New England, Armidale, NSW 2350, Australia; amanda.doughty@une.edu.au

This pilot study investigated how the movement order of a flock of sheep in an extensive environment differed over time to determine if order, as recorded by radio frequency identification tags, was stable and if deviations in position might be used to identify animals with compromised welfare state. The hypothesis was that flock 'free' movement order would be relatively stable and repeatable. Two hundred mature Merino ewes were trained to walk a distance of 1 km following a handler carrying a bucket of grain. The sheep were allowed to move at their own pace and, provided they were not grazing, were not pushed from the rear. Thirteen runs around a fenced track occurred over an eight week period during mid-pregnancy. General health (lameness, demeanour, posture) were recorded at each run, a blood sample to assess various haematological parameters (e.g. RBC, HCT, WBC, neutrophil, lymphocyte and eosinophil concentrations) was taken from each ewe between the 12th and 13th runs and a temperament test was conducted at the completion of the runs. The results indicated that the position of each ewe on each run varied greatly, except for the lead sheep who was always either in first or second position. The coefficient of variation (CV) was calculated for the position of each ewe for all the runs. This was then ranked into blocks of 20 ewes (lowest CV, mid CV, highest CV). Kendall's tau coefficient gave a value of -0.77 when correlating the CV against the mean position of each ewe, indicating that ewes at the back of the flock showed significantly less variation in position to those at the front ($P<0.001$). There were no correlations between CV and any of the blood parameters measured or temperament scores. These results indicate that movement order is more stable for animals normally found towards the back of the flock. Therefore, it is possible the change in movement order is not likely to be useful to identify risk of welfare compromise and, given the stability of the latter third of the flock it is likely that additional data will be needed to differentiate their health status.

Moving Geographic Information Systems research indoors: spatiotemporal analysis of agricultural animals

Courtney L. Daigle[1], Debsmit Banerjee[2], Robert A Montgomery[3], Subir Biswas[2] and Janice M Siegford[1]
[1]Mich. St. Univ., Animal Science, East Lansing, MI 48824, USA, [2]Mich. St. Univ., Electrical and Computer Engineering, East Lansing, MI 48824, USA, [3]Mich. St. Univ., Fisheries and Wildlife, East Lansing, MI 48824, USA; siegford@msu.edu

Wildlife ecology studies use Geographic Information Systems (GIS) to assess animal movement and behavior with unlimited space and finite resources. A GIS-approach for studying agricultural animal behavior in an indoor environment with finite space and unlimited resources was developed using laying hens as our animal model. Our objective was to demonstrate the potential for using GIS to provide information on where individual laying hens spent time in a non-cage environment and where they chose to engage in specific behaviors. Because this was a proof of concept using only 2 rooms, only descriptive statistics are presented. However, multiple hens per room were studied to try to assess response diversity among hens and within hens over time as a first step to monitoring using this technique at the level of the individual hen. Data collected concurrently from a wireless body-worn location tracking sensor system and video-recording examined spatially-explicit behavior from two identical indoor rooms (Rooms B and Y) of hens (135 hens/room). We spatiotemporally represented these data in GIS. Utilization distributions and hotspot maps depicted home range patterns and the spatial configuration of specific behaviors (feeding, foraging, and preening) at two ages (48 and 66 wk). Variation in home ranges (HR), HR overlap, and behavioral distribution show that individual hens respond to the same environment differently. For example Y10 used 11 and 12% of the room at 48 wk and 66 wk, respectively. Hen B2 used 5% of the room at 48 wk and 27% of the room at 66 wk. Hens B8 and B10 had the largest HR overlap of 6.21 m^2 at 48 wk and preened in similar locations, suggesting a possible social relationship. B10 used a smaller proportion of the room at 66 wk (6%) than at 48 wk (41%) and spent less time preening at 66 wk (1%) compared to 48 wk (19%), suggesting a welfare concern. Variation in the spatial configuration of hen foraging was also observed. Some hens exhibited an inverse relationship between feeding and foraging. Hen B1 spent 40% of her time feeding and 16% foraging; hen B8 spent 1% of her time feeding and 22% foraging, suggesting individual hens employ different strategies for acquiring food. The genetically-similar hens studied here exhibited diverse behavioral and spatial patterns when examined using GIS. This approach can provide detailed examinations of individual non-cage laying hen behavior. Such an approach could be a powerful research tool and is applicable to other agricultural animals in stochastic indoor environments.

Assessment of cage space in breeding mice

Brianna N. Gaskill[1,2] and Kathleen R. Pritchett-Corning[2,3]
[1]Purdue University, Comparative Pathobiology, 625 Harrison St., West Lafayette, IN 47907,
USA, [2]Charles River Laboratories, 251 Ballardvale St, Wilmington, MA 01887, USA, [3]Harvard
University, Office of Animal Resources, 16 Divinity Ave, Cambridge, MA 02138, USA;
bgaskill@purdue.edu

Recommendations for the amount of cage space for breeding rodents were recently increased in the 2011 Guide for the Care and Use of Laboratory Animals (USA). The few studies conducted on cage space for breeders utilized custom built caging, making practical application difficult. We hypothesized that if a difference in behavior and reproduction exists within the limits of commercially available caging, this difference would be detected between the smallest and largest cages. C57BL/6 and CD-1 breeding mice were assigned to a cage treatment: LP-18790 (226 cm^2); A-RC1 (305 cm^2); A-N10 (432 cm^2); T-1291 (800 cm^2) and a breeding configuration: Single (Male removed after birth); Pair (Male+Female); Trio (Male+2Females) in a factorial design for 12 weeks. All cages received 8 g of nesting material and nests were scored weekly. Pups were weighed and sexed at weaning. Adult behavior and location in the cage were recorded by scan samples every 30 min over 48 hr of video (weeks 1 and 3 of pups in the cage). Corner inactivity and play behavior were recorded by 1/0 sampling method. All data was analyzed using a GLM. Cage space did not significantly alter any reproductive measures. Mice in LP-18790 were inactive while in contact with their pups more often than those in T-1291 ($P<0.05$). Pups in the LP-18790 cage played less than in the other cages ($P<0.05$). Adults in LP-18790 displayed more corner inactivity than the two largest cages ($P<0.05$). Only mice in cage T-1291 spent more time under the feeder than in other areas of the cage ($P<0.05$). Nests scored the highest in T-1291 ($P<0.05$). Housing breeding mice in a range of commercially available cage sizes does not affect reproduction but behavioral measures suggest that LP-18790 is mildly stressful. Based on the measure of corner inactivity, adults may benefit more from complexity that allows for escape from pups instead of additional floor space.

Development of an exploratory test and a contrast test for mice
Anette Wichman and Elin Spangenberg
SLU, Animal Environment and Health, Box 7068, 750 07 Uppsala, Sweden; anette.wichman@slu.se

A future aim in our project is to be able to assess positive emotional states in mice based on their behaviour. However, many of the standardized behavioural tests for mice use aversive components to assess the degree of anxiety in relation to exploration. Therefore we set out to design an exploration test and a contrast test that would be rewarding for the mice and give indications of how positive they experience the situation. We used 51 C57BL/6N mice, 24 females and 27 males, age 20 weeks and housed in single sex groups of three in Macrolon 3 cages with wood-shavings, cardboard house, cellstoff paper and ad lib access to food and water. The exploration arena (100×100 cm) was divided into nine sections which contained different items such as hay, cardboard rolls and small branches. The mice entered the arena from a start box, placed in the middle, either by passing a push-door with increased resistance (motivation) or there was an increasing time delay for the start box door to open automatically (anticipation). Behaviour in start box and number of visits to the different sections during 7 minutes in the arena were scored. In the contrast test the mice traversed a runway (140×12 cm) where half the mice received a large reward (hazelnut crème) and the other half a small reward (food pellet) at the far end. Latency to reach the reward was recorded. Baseline treatments were run for 7 days (3 trials/day, trial 1-21) and then rewards were shifted for half of the mice from each category, to establish a negative and positive contrast. The contrast treatments were run for 3 days (3 trials/day, trial 22-30). The maximum weight of the push–door the mice passed was 105.4±16.07 g (mean±SD) (mean body weight 26.7±2.7 g). Anticipation mice performed more behaviour transitions (Mann-Whitney U test; $P=0.001$) and sniffed more on the door ($P=0.012$) in the start box compared to control mice that only spent time in the start box. Comparing latency between baseline trial 16-21 and trial 23-27 in the contrast phase, the mice experiencing a positive contrast decreased their latency to reach the reward (Wilcoxon test; $P=0.028$) but for negative contrast there was no significant difference. After the shift there was no significant difference in run time between positive contrast and the control treatment that continued on the large reward. The results imply that the mice were positive towards entering and exploring the exploratory arena as they worked to get through the push-door and showed indications of anticipation. The contrast test also showed indications of a positive response with increased reward, although no real contrast effects between the treatments were detected. Thus, the tests have shown potential for further use into investigations of positive experiences in mice.

Utilization of nesting material by laboratory hamsters

Christina Winnicker and Brianna Gaskill
Charles River Laboratories, Animal Welfare & Training, 251 Ballardvale St., Wilmington, MA
01887, USA; christina.winnicker@crl.com

Like other rodents, hamsters build nests during late gestation through the birth and nursing of pups. While extensive assessment of nesting material has been done with mice, similar work has not been previously done for hamsters. We previously assessed multiple commercially available materials and amounts and selected two that resulted in the highest scoring nests. Our hypothesis was that when provided with behaviorally appropriate nesting material hamsters would build significantly better quality nests than those housed on bedding alone, and the nest would provide a thermoregulatory advantage leading to increased litter sizes, increased pup survival to weaning and increased pup weaning weights. Twenty grams of two different nesting material treatments (tissues or shredded paper) were provided to late-term pregnant hamsters through weaning of their pups in addition to the standard aspen shaving bedding (controls). Nests were scored weekly with a previously established scale. Data was analyzed as a GLM with post hoc Tukey tests. Both nesting materials (P<0.05) showed significantly improved nest scores over control cages with bedding alone. No difference was seen in the number of pups born ($F_{2,12}$=0.3158, P=0.735) or the weaning weights ($F_{2,10}$=0.7516, P=0.496) for any of the treatments. Pups born in cages with shredded paper (P<0.05) nesting material had less mortality than tissue nested cages (P<0.05) but was indifferent from controls (P>0.05). Thus, 20 grams of either tissue or shredded paper make a suitable nesting material for building in laboratory hamsters, but the shredded paper alone improved hamster welfare by improving pup survival to weaning.

Play in rats: association across contexts and types, and analysis of structure

Luca Melotti[1], Jeremy D. Bailoo[1], Eimear Murphy[2], Oliver Burman[3] and Hanno Würbel[1]
[1]Division of Animal Welfare, University of Bern, Länggassstrasse 120, 3012 Bern, Switzerland, [2]Emotion and Cognition Group, Utrecht University, Department of Farm Animal Health, Yalelaan 7, 3584 CL Utrecht, the Netherlands, [3]School of Life Sciences, University of Lincoln, Riseholme Park, LN2 2LG Lincoln, United Kingdom; luca.melotti@vetsuisse.unibe.ch

Play is proposed as a promising indicator of positive animal welfare among higher vertebrates. We aimed to study play behaviour in rats across contexts (conspecific/heterospecific) and types (social: pinning, being pinned; solitary: scampering), and we investigated its structure using behavioural sequence analysis. Group-housed (three per cage) adolescent male Lister Hooded rats (age: 6-7 weeks, n=21) were subjected to a Play-In-Pairs Test: after a 3.5 hour isolation period, a pair of cage-mates was returned to the home cage (area × height: 1,820 cm^2 × 20 cm) and both social and solitary play were scored for 20 minutes. This procedure was repeated for each pair combination across three consecutive days, and scores from each combination were combined to form individual play scores. Heterospecific play was measured using a Tickling Test: after habituation, rats were individually tickled by the experimenter in a test arena through bouts of gentle, rapid finger movements on their underside, during which number of positive 50 kHz frequency modulated vocalisations were recorded. Both tests above were compared with social play in the home cage during the first hour of the dark cycle. While conspecific play in both the Play-In-Pairs Test and home cage were correlated (r(18)=0.57, P=0.008, controlling for cage), both were unrelated to heterospecific play in the Tickling Test (positive vocalisations; r_s=-.07, n.s., and r_s=0.28, n.s., respectively). During the Play-In-Pairs Test, although both solitary and social play types occurred, they were unrelated (r(18)=0.16, n.s., controlling for cage), and solitary play of one rat did not predict the subsequent play behaviour of its cage mate (F(2,40)=0.27, n.s.). Behavioural sequence analyses of play in the Play-In-Pairs Test revealed that within each rat, repetitions of either pinning or being pinned behaviour were more frequent than alternations between them (t(20)=9.89, P<0.001), while this difference was not found for scampering behaviour (t(20)=-0.18, n.s.). These results suggest that play behaviours in conspecific and heterospecific contexts are independent, and that social and solitary play are unrelated in rats. Analysis of play structure revealed that social play occurs in bouts of repeated behaviours while solitary play sequences do not follow a specific pattern. If play is to be used as an indicator of positive welfare in rats, context, type and structure differences should be taken into account.

Latency to start interaction after working for the resource in blue foxes

Tarja Koistinen, Elina Hämäläinen and Jaakko Mononen
University of Eastern-Finland, Department of Biology, P.O. Box 1627, 70211 Kuopio, Finland;
tarja.koistinen@uef.fi

In the studies measuring the strength of animals' behavioural motivations, the entrance to the resource is often made costly, not the actual interaction with the resource. This allows the animals to schedule their interaction with the resource: for example, the interaction may be delayed or the resource is left unutilised. Here we compare the latencies to start interaction with various resources in farmed blue foxes. Ten juvenile blue fox males were housed individually in experimental apparatuses where they could enter a resource, i.e. a wooden block (WB), nest box (NB), platform (PL) or sand floor (SF), one at a time, by opening a push-door with 3.5 kg extra weight. The first interaction (behaviour) with the resource and the latency for the first interaction were analysed from a total of 474 entrances through the push-door. The data was analysed by using Linear Mixed Model (SPSS). Interaction with the resource (within 10 mins) after an entrance was observed more often in the case of NB (in 97±5% of entrances, mean±SEM) and SF (96±5%) than WB (60±5%) and PL (54±5%) ($F_{3,34}=21.8$, $P<0.001$). The latency to start interaction was shortest for SF (8±1 sec) and longest for PL (113±16), WB (64±12) and NB (25±7) being in between these two ($F_{3,335}=26.3$, $P<0.001$). The latency to interact with the resource depended on the first observed behaviour in the case of WB and NB ($F_{11,334}=3.0$, $P<0.01$), but not in the case of SF and PL. For WB the latency was shortest for vole jumping and play and longest for pawing and poking. For NB the latency was shortest for observation on the roof of the nest box and longest for stretching against the nest box. Overall, investigation (sniffing, licking) was the most common first interaction with all the resources (57% of cases) before the more resource-specific behaviours. In conclusion, the results indicate that the vicinity of the resource may be important per se, and not only the interaction with the resource. The urge to interact with the resource after working for it depends on the type of the resource, suggesting following order of importance in farmed blue foxes: SF>NB>WB>PL. Initial investigation seems important before other types of interactions with a familiar resource, indicating a fixed order of activities with the resources.

Is the more the merrier true for young mink?

Lina Olofsson[1], Steffen W Hansen[2] and Lena Lidfors[1]
[1]Swedish university of agricultural sciences, Department of animal environment and health, Gråbrödragatan 19, 532 31 Skara, Sweden, [2]Aarhus university, Department of animal health, welfare and nutrition, Blichers alle 20, 8830 Tjele, Denmark; lina.olofsson@slu.se

Climbing cages for farmed mink will be the rule in Sweden from 2017, but there is an ongoing discussion about the effect on animal welfare of keeping pairs or groups of three mink during the growing period. Hence, group housing may reduce fur chewing, increase bite marks, give uneven weight gain and reduce the use of the nest box for some animals. The aim of this study was to compare differences in behaviour and physical measurements between mink held in pairs or in groups of three in climbing cages. The study was conducted from July to November on a private farm in brown mink kept in pairs (P: one male, one female, n=12 cages) or in groups (G: one male, two females, n=12 cages). Cages consisted of a lower (0.276 m^2) and upper floor (0.138 m^2), a nest box (0.015 m^3) with saw dust and straw, a shelf on each floor and a chain as enrichment. There were ad libitum access to water and fresh food. Three video recordings were done during September to November for 24 h every fourth week. Videos were analysed with scan-sampling every five minutes. All mink where weighed in July, September and October. After euthanasia and pelting in November recordings were done of fur chewing, bite damages on the fur, and bite marks on the leather side of the skin. Statistical analyses were done with Wilcoxon Rank Sum test on behaviours, Students T-test on weights and Fisher's exact test on fur chewing, bit damages and bite marks. Results showed differences in active behaviour between pairs or groups of mink (median P: 19.4%, G: 20.4%, P=0.49). Pairs used the nest box more (median P: 79.4%, G: 69.2%, P<0.01) and groups rested more with body part outside the nest box (median P: 0.3%, G: 3.3% P<0.01). There were no differences in weight in September between pairs and groups for males (mean P: 2.91 ± 0.067 G: 2.86 ± 0.080 P=0.58) or females (mean P: 1.69 ± 0.054 G: 1.68 ± 0.026 P=0.81) neither in October for males (mean P: 3.39 ± 0.079 G: 3.24 ± 0.089 P=0.24) or females (mean P: 1.84 ± 0.034 G: 2.04 ± 0.216 P=0.38). There were no differences in fur chewing (P: 27.12% G: 26.97% P=1.0) or bite damages on the fur (P: 1.7% G: 6.7% P=0.24) between groups and pairs. There were higher occurrences of bit marks on the leather side in groups than in pairs for both black old marks (P: 17.6% G: 41.2% P<0.01) and red fresh marks (P: 4.1% G: 15.5% P<0.05). In conclusion, the more the merrier can be questioned if considering that the presence of bite marks would be negative for mink, but there were no effects on the other measures and the difference in nest box use could be due to a difference in temperature between the days of recording the two treatments.

Investigating anhedonia in a non-conventional species: are some riding horses depressed?

Carole Fureix[1], Cleo Beaulieu[1], Céline Rochais[2], Soizic Argaud[2], Séverine Henry[2], Margaret Quinton[1], Martine Hausberger[2] and Georgia Mason[1]
[1]University of Guelph, Animal Poultry Science, 50 Stone Road East, N1G2W1 Guelph, Canada, [2]Université Rennes 1, UMR CNRS 6552, Campus de Beaulieu, 35042 Rennes Cedex, France; cfureix@uoguelph.ca

Some riding horses display states of inactivity and low responsiveness to external stimuli that we term 'withdrawn', and that resemble the reduced engagement with the environment seen in clinically depressed people. To assess whether these animals are indeed affected by a depression-like condition, we investigated anhedonia: the loss of pleasure that is a core symptom of human clinical depression. Subjects were withdrawn horses and controls from the same stable (16 geldings and 4 mares, 7-20 years old, 85% French Saddlebred). The time individuals spent being withdrawn was determined by a trained observer using instantaneous scan sampling every 2 minutes over 1 h long periods repeated daily over 15 days. To measure sucrose intake, a classic measure of anhedonia in rodent-based biomedical research never previously applied to horses, commercially-available flavoured sugar blocks, novel to these subjects, were mounted in each stall and weighed 3, , 24 and 30 h after provision. We hypothesized that if depressed-like, withdrawn horses would consume less sucrose than controls. Horses spending the most time withdrawn did show reduced sucrose consumption ($F_{1,18}$=4.65, P=0.04, in a repeated measures model also controlling for age, sex, and the time each horse spent in its stall – thus able to eat the sucrose – during testing). We then controlled for two possible alternative explanations for this pattern: neophobia towards novel foods, and generally low appetites. Hay consumption was measured over 5 days, as were subjects' latencies to eat a meal scented with a novel odour. When included in our model, high hay consumption strongly tended to predict high sucrose consumption ($F_{1,14}$=4.52, P=0.051), while long latencies to eat a novel food predicted low sucrose consumption ($F_{1,14}$=8.34, P=0.012). However, statistically controlling for these two confounds did not eliminate the relationship between being withdrawn and consuming less sucrose (although reducing it to a strong trend: $F_{1,15}$=4.28, P=0.056), suggesting that neither overall food consumption levels nor neophobic reactions explained our previous findings. Overall, this study illustrates the methodological challenges of investigating anhedonia in non-conventional species; reveals possible depression-like conditions in riding horses; and suggests a way of assess anhedonia in other animals showing profound inactivity (e.g. working equids in the developing world).

The sow's udder: a neonatal piglet's foraging territory, battlefield and 'dance floor'

Janko Skok and Dejan Škorjanc
University of Maribor, Faculty of Agriculture and Life Sciences, Department of Animal Sceince, Pivola 10, 2311 Hoče, Slovenia; janko.skok@um.si

After birth, neonatal piglets are exposed to a challenging social and physical environment. Littermates share the same 'foraging' territory, i.e. the sow's udder. During the lactation period, piglets experience intense social (including aggressive) interactions with their littermates until they establish a reliable teat order on the mother's udder. We revealed that in the first two weeks after farrowing, when piglets appear randomly distributed on the udder, geometric constraints (suckling space limitations) are important force driving the piglets' distribution along the limited udder. An effect of geometric constraints results in the specific hump-shaped distribution, known as the mid-domain effect. In addition, our results show that prior to the establishment of teat order, piglets rely on group suckling cohesion, an order mechanism that refers to the maintenance of significantly similar inter-individual distances on the udder. Although group suckling cohesion operates, neonatal piglets are still confronted with a direct struggle with their littermates. Despite the possible adaptive function of fighting in competition for limited colostrum, a certain level of suckling orderliness is essential because it reduces exaggerated teat disputes and the risk of missing suckling, which consequently affects survival. Suckling cohesion occurs first, after which the teat order develops (piglets fix their positions at the udder). The mechanism underlying the choice that leads each piglet to the same suckling position is still not clarified. However, our observations indicate that the spatial configuration of the sow's udder, determined by two parallel rows of teats arranged in pairs, can hypothetically be proposed as an important cue for the suckling orientation of piglets. In conclusion, our studies elucidate some important aspects of the suckling behaviour in piglets, revealing complex littermates' interactions, which are also subject to the spatial features of the udder.

Mother-infant social interactions in rangeland-raised beef cattle

Mohammed N. Sawalhah, Andrés F. Cibils and Vanessa J. Prileson
New Mexico State University, Animal and Range Sciences, MSC 3I, P.O. Box 30003, Las Cruces, NM 88005, USA; acibils@nmsu.edu

The objective of this study was to document free ranging cow-calf proximity patterns to assess whether maternal-induction of foraging behaviors is likely to be important in free ranging beef calves. Location data collected on 28 cross-bred cows and 14 calves that grazed two adjacent 219 and 146 ha pastures were used to determine the spatial and temporal relationships among range calves, their dams, and other nursing cows in the herd at two different calf ages (1 or 4 months). Cow and calf positions were recorded with GPS collars at 5-min intervals during 14 d over four grazing periods between 2009 and 2011. Association software was used to analyze GPS data and develop association matrices including all cows and calves at two different spatial thresholds (5 and 10 m). Social structure software was used to conduct cluster analysis of the association matrices using the average association option. Cows other than the dam that spent the most time in the proximity of the focal calf were considered 'babysitter' cows. Different cows performed the 'babysitter' function for a given calf on different days. Calves spent on average 11 and 21.7 h/day within 5 or 10 m of an adult cow, respectively, and spent 2.9, 8.1, and 2.1 h/day within 5 m of their dams, other adult cows, or a 'babysitter' cow, respectively. Calves spent significantly ($P<0.01$) more time within 5 m of other cows (8.1 h/day) vs. their dam (2.9 h/day) regardless of age, and spent a similar ($P=0.77$) amount of time within 5 m of their dam vs. the 'babysitter' cow at 1 month (1.8 vs. 2.0) but not at 4 months of age (4.2 vs. 2.1; $P=0.02$). Calves spent more time within 5 and 10 m of their dams at 4 vs. 1 month of age (4.2 vs. 1.8 and 7.3 vs. 3.2 h/day, respectively). Cluster analyses showed higher mother-infant clustering at 4 vs. 1 month of age regardless of the spatial threshold considered. Calves spent relatively little time in the proximity of their dams, especially at a young age, and tended to have more opportunities for interaction with other nursing cows in the herd. If early life learning of foraging behaviors occurs through imitation and is a function of the time a calf spends close to an adult cow, then the influence of the dam vs. other nursing cows may be less important than previously believed.

Validation of HOBO Pendant® accelerometers for automated step detection in adult male turkeys

Hillary A. Dalton[1], Benjamin J. Wood[1,2], James P. Dickey[3] and Stephanie Torrey[1,4]
[1]*University of Guelph, Department of Animal and Poultry Science, 50 Stone Road East, Guelph, Ontario, N1G 2W1, Canada,* [2]*Hybrid Turkeys, Suite C, 650 Riverbend Drive, Kitchener, Ontario, N2K 3S2, Canada,* [3]*Western University, School of Kinesiology, 3M Centre, Room 2225, London, Ontario, N6A 5B9, Canada,* [4]*Agriculture and Agri-Food Canada, 93 Stone Road West, Guelph, Ontario, N1G 5C9, Canada; hdalton@uoguelph.ca*

Activity levels can be used as a predictor of factors such as health status and temperament in a number of different farm animal species. However, video or live behavioural observations of activity levels are often time consuming or unfeasible in very large groups as are standard housing for commercial turkeys. Small, portable data loggers have been recently validated as a cheap and effective alternative to video or live behavioural observation for the automated detection of postural and stepping activity in several livestock species. We aimed to validate the use of HOBO Pendant® data loggers for automated step detection in adult male turkeys housed in free-run pens. Eight 13-wk old Hybrid Convertor tom turkeys were habituated to HOBO® Pendant G Acceleration Data Loggers fastened to one of their hocks with a Velcro®-sealed white elastic bandage for a 3-hr period daily for 2 wk. During the one day data collection, the accelerometers were programmed to record each bird steps as x- and y- axis acceleration readings every 0.01 s over 54 min. Simultaneous video observation allowed the logger data to be evaluated for sensitivity and accuracy in reporting walking activity for these adult toms. The accelerometer data were transformed with a derivation of the Pan-Tompkins algorithm. Steps were identified by thresholding the derivative of the resultant x- and y-axis band-pass filtered accelerations. Peaks above arbitrary threshold of 0.55 g/s were classified as steps. The acceleration data and video observations were divided into 2-minute blocks for comparison of steps taken. The results showed a relationship of $r^2=0.747$ across all eight toms and suggest HOBO data loggers have the potential to be used to monitor adult turkey activity.

Use of a bird-mounted tri-axial accelerometer to quantify experienced physical forces during descent in laying hens

Janja Širovnik[1], Francesca Booth[2], John F. Tarlton[2] and Michael J. Toscano[1]
[1]University of Bern, ZTHZ, Division of animal welfare, Burgerweg 22, 3052 Zollikofen, Switzerland, [2]University of Bristol, Matrix Biology Research Group, Langford, BS8 1TH Bristol, United Kingdom; janja.sirovnik@vetsuisse.unibe.ch

Keel bone fractures are a major problem in laying hens and were recently suggested by the Farm Animal Welfare Council as the greatest issue facing commercial producers. Nonetheless, there is still a considerable lack of knowledge as to the cause of keel fractures. To provide a greater understanding of the physical forces involved in causing keel damage, we have developed a tri-axial accelerometer fitted within a custom-designed vest that is worn by hens during locomotion to record acceleration at 1000 Hz. The sensor is positioned directly on the keel and thus is able to provide an accurate representation of physical forces the keel experiences. Our expectation was that the force on the keel bone of hens increases with the height of descent. For the study, six laying hens were equipped with accelerometer devices and trained to descend from several perch heights (40, 60, 90, and 130 cm) to the ground to access a food reward. With each descent, the device recorded the associated 3D acceleration from which the resulting output was examined for the maximal acceleration. Kinetic energy of descent was also calculated from the area under the curve of the acceleration vs time profile. Collected data were then statistically modelled to determine the relationship between the outputs (area under the curve, maximal acceleration) and descent height. Our analysis identified a positive linear correlation between the area under the acceleration curve (P=0.03) and height height of descent. We also identified a statistical tendency between the maximal acceleration and the height of descent (P=0.069). Our results demonstrated that changes in velocity are correlated with the height of descent, but the correlation between the maximal acceleration and the height is less clear, possibly because the larger height provides hens more time to prepare for landing, i.e. wing-flapping. Future experiments are needed to quantify the benefit of related behaviours in reducing physical forces on the keel. We believe that researching the biomechanical forces involved in descent and assessing the keel bone condition will contribute to understanding the causes of fractures.

Personality traits of high, low and non users of a free range area in laying hens

Carlos E. Hernandez[1,2,3], Caroline Lee[2], Drewe Ferguson[2], Tim Dyall[2], Sue Belson[2], Jim Lea[2] and Geoff Hinch[1,3]
[1]*University of New England, School of Environmental and Rural Science, Armidale, 2351, Australia,* [2]*CSIRO, Animal, Food and Health Sciences, Locked Bag 1, Armidale, NSW, 2350, Australia,* [3]*Poultry CRC, Armidale, NSW, 2351, Australia; chernand@une.edu.au*

Free range poultry production is often perceived as a welfare friendly production system since it provides an outdoor range (OR) that allows the birds to express their full range of natural behaviours. However, a significant proportion of birds never use the OR and this could partly be due to individual characteristics of the animals. The aim of this study was to characterize personality traits of laying hens with different levels of OR use. We hypothesized that non-users of the OR would display higher levels of fear and stress than OR users. Two hundred ISA brown hens were individually tagged with RFID plastic leg-bands. Hen movements between an indoor and outdoor area were monitored daily from 9:30 am to 4 pm, between 22 and 31 weeks of age, using RFID antennas fitted to the pop-holes. At 32 weeks of age, 60 birds were selected for personality trait tests based on their OR usage: 20 birds with High OR usage (H-OR; visiting OR daily), 20 birds never using OR (No-OR) and 20 birds with Low OR usage (L-OR; visiting OR<5 days). Personality trait tests included: 5 min manual restraint (MR; struggles and vocalizations, plasma corticosterone levels: basal and 10 min post-restraint), 5 min open field (OF; vocalizations and activity – assessed as number of squares crossed inside arena) and tonic immobility test (TI; attempts to induce and duration of immobility). Data were analysed using standard least squares analysis with post hoc LS Mean contrasts or t-test. We found that No-OR birds responded to MR with a significant increase in plasma corticosterone levels (basal=1.75±0.27 vs 10 min=2.92±0.27 ng/ml, P<0.01) whereas L-OR and H-OR birds did not display such response (L-OR: basal=2.14±0.40 vs 10 min=2.44±0.38 ng/ml and H-OR: basal=2.10±0.33 vs 10 min=2.35±0.33 ng/ml, both P=0.6). In the OF test, No-OR birds tended to be slower to first move (No-OR=148.3±26.8 s vs H-OR=87.7±25.4 s, P=0.1) and to walk less (No-OR=38.2±13.9 vs H-OR=65.8±13.2, P=0.1) than H-OR, but not L-OR hens. Furthermore, three No-OR birds displayed 'freezing' behaviour while this was not observed in H-OR and L-OR birds. During TI test, No-OR birds tended to take longer to stand up (No-OR=154.7±19.5, L-OR=88.9±19.5 and H-OR=107.8±20.0 s, P=0.1). In conclusion, non-users of the outdoor range showed a higher corticosterone response to an acute stress challenge and tended to be more fearful when separated from their social group and exposed to novel environments. Personality traits may explain the level of use of the outdoor area in laying hens.

Analysis of movement in pullets: benefits of larger groups

Guiomar Liste[1], Irene Campderrich[1] and Inma Estevez[1,2]
[1]Neiker-Tecnalia, Department of Animal Production. P.O. Box 46, 01080 Vitoria, Spain,
[2]Ikerbasque, Basque Foundation for Science. Alameda Urquijo 36-5 Plaza Bizkaia, 48011 Bilbao,
Spain; mguiomar@neiker.net

Alternative housing for laying hens, such as enriched cages, offer wider behavioural opportunities but it might also increase the risk of undesired behaviours, with important consequences for the birds' health and welfare. However, larger groups could also translate into larger enclosures and increased space efficiency at constant densities, which could improve movement and activity opportunities. This experiment aimed to analyze the effects of group size (GS), phenotypic appearance diversity (PA) and age on movement trajectories of pullets. 1050 Hy-line Brown chicks were divided into 45 experimental pens housing 10, 20 or 40 birds at a constant density of 8 hens/m². Simultaneously, different proportions of modified PA treatments were studied in a full factorial set up (0, 30, 50, 70 or 100% of altered birds per group) by placing a black mark at the back of the chicks head. Hence, for each GS and PA combination there were 3 replications with all GS populations being either homogenous (100 or 0%), with marked (M) or unmarked (U) phenotypes, or heterogeneous (30M/70U, 50M/50U or 70M/30U) with both phenotypes coexisting in the same pen. Birds were observed at 3 age periods during the rearing phase: 5-6 weeks (P1), 10-11 weeks (P2) and 15-16 weeks (P3). The software Chickitizer was used to record locations (XY coordinates scored every 10 sec). Total, net, maximum and minimum distances travelled were calculated from the coordinates, as well as angular dispersion index and net to total distance ratio to estimate the sinuosity of the trajectories. GLMM models considered GS and PA as fixed factors, age as repeated measure and pen as random effect. Total, net and maximum distances were affected by GS ($P<0.001$, all comparisons) with birds in GS 40 travelling the longest total, net and maximum distances (208.0 ± 6.8, 70.3 ± 2.7, 56.3 ± 1.8 cm respectively; $P<0.05$, all comparisons). Angular dispersion index was higher in groups of 10 than 20 ($P<0.01$). Total, net, maximum and minimum distances declined over time ($P<0.05$, all comparisons), but net to total distance ratios were higher at P3 than P1 ($P<0.05$). Results suggest that, despite the constant densities, larger GS translated into longer total and net distances travelled as total available space was greater. PA did not affect the way pullets moved through the space probably due to the early marking and the long-term social stability of the groups. Trajectories shortened with time which could be explained by the larger size and general decline in activity observed in older birds. In conclusion, birds housed in larger GS at constant density may benefit from better space efficiency which results in higher bird mobility.

Does LED-lighting in broiler houses improve welfare?

Anja Brinch Riber
Aarhus University, Department of Animal Science, Blichers Allé 20, 8830 Tjele, Denmark;
anja.riber@agrsci.dk

Broiler houses are mainly lit by fluorescent light (FL). Light-emitting-diodes (LED) are an energy saving alternative that are more durable and retain the light intensity for considerably longer periods. Therefore, we hypothesised that LED-lighting generates a more uniform distribution of light intensity, causing an improved spatial distribution of broilers that results in positive effects on welfare. Two identical broiler houses were equipped with FL and LED, respectively. Data were collected from six flocks/house of 23,300 Ross 308. Light intensity was measured on days -1 and 34 in 40 positions evenly distributed in the house. To estimate the spatial distribution of broilers, six cameras/house each photographed an area of known size every second hour/fifth day. Distributions of light and broilers were analysed using the standard deviations as response variables. On day 34, 200 broilers/flock were assessed for gait and contact dermatitis. All data were subjected to repeated measurements analysis using flock as the statistical unit. On day -1 a more uniform distribution of light intensity was found in FL, whereas no difference was found on day 34 (day*treatment: $F_{1,11}=7.58$; P=0.02). The spatial distribution of broilers did not differ between the two treatments ($F_{1,269}=0.96$; P=0.33). When differences were found in gait score, foot pad dermatitis, and hock burns, better scores were found in LED (treatment × age of light source: $9.93<F_{10,2388}<30.27$; P<0.001). The use of LED-lighting did not result in a more uniform spatial distribution of light intensity and hence broilers. Although our hypothesis was not supported, we found improved welfare of broilers reared with LED-lighting. We speculate whether this finding is caused by a slower initial growth of broilers reared with LED-lighting allowing bones to grow stronger before weight is put on. Further studies are needed to elucidate the causes of the improvement found in welfare.

Factors influencing the inter-group agonistic behaviour of free-ranging domestic dogs (*Canis familiaris*)

Sunil Kr Pal

Katwa Bharati Bhaban School, Life Science, Katwa Abasan, MIG (U), 68, P.O., Khajurdihi, Dist., Burdwan, 713518, India; drskpal@rediffmail.com

The purpose of this study was to investigate the factors influencing agonistic behaviour of free-ranging domestic dogs (Canis familiaris). Observations on the inter-group agonistic behaviour of 21 free-ranging dogs from four neighbouring groups were recorded in Katwa town, India. Observations on dogs' behaviour were conducted daily along the fixed itineraries; and a total of 2880 h over 360 days was devoted to collect data. Behavioural data were collected using ad libitum and focal animal sampling. Group home range size was maximum during the late monsoon months, and then during the winter months. Seasonal home range size was greater in the case of male dogs than in the case of female dogs (t=3.44, d.f.=19, P<0.0027). Adult dogs displayed a higher rate of agonism than juvenile dogs. Although, there were no significant differences between male and female dogs in relation to their agonism, the frequency of inter-group aggressive encounters was highest during the late monsoon months whereas the frequency of inter-group submissive encounters was highest during the winter months. Male dogs were observed to initiate more agonistic encounters at mating places and also at territorial boundaries, whereas female dogs were observed to initiate more agonistic encounters at feeding places and also in the vicinity of den areas, showing the evidence of context dependent agonism (F=25.79; d.f.=4, 80; P<0.0001). Interestingly, aggressive postures displayed by the dogs varied with the competitive contexts, although the variation in the case of submissive postures being not statistically significant. Most aggressive postures were displayed by 'barking, growling and snarling' and most submissive postures were displayed by 'lips retracted in a submissive grin'. From this study, it may be presumed that different intrinsic and extrinsic factors may have different effects on the aggressiveness of the free-ranging dogs.

Heritability of human directed social behaviours in beagles

Mia Persson, Lina Roth, Martin Johnsson and Per Jensen
IFM Biology, Division of Zoology, AVIAN Behavioural Genomics and Physiology Group, Linköping University, 581 83 Linköping, Sweden; mia.persson@liu.se

Through domestication and co-evolution with humans, dogs have developed certain human-like social skills such as sensitivity to human ostensive cues and comprehension of referential gestures. However, dogs have also evolved abilities to attract human attention e.g. in a manner of asking for help when faced with a problem solving task. The aims of this study were to investigate within breed variation in human directed contact seeking behaviour in dogs and to estimate its genetic basis. To do this, 500 beagles bred and kept under standardized conditions at a research dog kennel, were tested once each in an insolvable problem task in the presence of an inactive test leader previously unknown to them, and their behaviour was video recorded. Durations, frequencies and latencies were recorded for interactions with the test equipment and the test leader. Contact seeking behaviours recorded involved both eye contact and physical interactions. Behavioural data was analysed with principal component analysis, resulting in four factors explaining 36, 20, 9.3 and 6.5% of variance respectively (factor 1: test equipment interactions; factor 2: social interactions; factor 3: eye contact; factor 4: physical interactions). There was no effect of age or sex on Factor 1 scores, while factor 2 scores differed significantly between males (-0.23, SEM 0.07) and females (0.19, SEM 0.07) (Univariate GLM, P<0.01) where females performed more social interactions than males. Factor 3 scores were significantly affected by age (P<0.01) where older dogs (-0.12, SEM 0.07) sought eye contact more often than younger dogs (0.12, SEM 0.06). Finally, the fourth factor involved behaviours related to physical contact and the scores differed significantly (P<0.05) between males (=-0.14, SEM 0.07) and females (=0.12, SEM 0.07) where females were more physical. Narrow sense heritability (h^2) and 95% highest probability density (HPD) interval was calculated for the principal components scores using the MCMCglmm animal model package in R. For factor 1 scores, h2 was estimated to 0.32 (0.13; 0.51) and for factor 2 scores, h^2 was estimated to 0.23 (0.06; 0.42). For the other factors the HPD interval was close to 0. These results show that within the studied dog population, behavioural variation in human directed social behaviours was sex dependent and that the utilization of eye contact seeking increased with either age and/or experience. Heritability estimates indicates a significant genetic contribution to the variation found in test equipment interactions as well as in social interactions.

Red Junglefowl (*Gallus gallus*) selected for low fear of humans are more dominant and show signs of improved welfare

Beatrix Agnvall, Anser Ali, Sara Olby and Per Jensen
Linköping University, IFM, Linköping University, 581 83 Linköping, Sweden; beaek@ifm.liu.se

Reduced fear of humans is a major component in the domestication of animals. It has been suggested that many of the traits associated with domestication could have developed as correlated responses to such reduced fear. Furthermore, it is possible that increased tameness may have reduced the stress of living in human proximity, and thereby improved the welfare and the reproductive potential (fitness) in captivity. To investigate this, Red Junglefowl (ancestors of domestic chickens) were selected for four generations on high or low fear response towards humans, generating three strains, High (H), Low (L) fear as well as an unselected (U) strain. We investigated possible correlated effects on social behaviour and social dominance, and changes in other, non-behavioural traits relating to welfare and fitness. Social behaviour and dominance was recorded in two different social tests. The first focused on the general social behaviour of the animals within strains (n=99, 54 males and 45 females, 33 animals from each strain) and the second on social dominance between strains when an attractive resource was limited, meal worm (n=99, 54 males and 45 females, 33 animals from each strain), water or dust bath material (n=42, 21 males and 21 females, 14 animals from each strain). Growth and plumage condition (n=101), as well as size of eggs and offspring (n=106) were also recorded. Birds from the L-line had higher weight (L=972±30, H=886±23, U=904±22; ANOVA $F_{(2, 98)}$=3.8, P=0.02), laid larger eggs (L=39±0.7, H=35±0.6 U=40±0.8; ANOVA $F_{(2, 14)}$=6.2, P=0.01) and generated larger offspring (L=28±0.2, H=25±0.3 U=29±0.3; ANOVA $F_{(2, 104)}$=37.6, P<0.001), and they had a better plumage condition (Scored from best to poorest on a scale 1-5; L=2.4±0.2, H=3.2±0.1, U=2.8±0.1; ANOVA $F_{(2, 95)}$=9.0, P<0.001). There were no significant differences between strains in the general social behaviour. However, between strains, L-birds performed more aggressive behaviour in the social dominance test regardless of whether the restricted resource was food or not; meal worm (L=8.06±1.2, H=3.21±1.3, U=2.72±1.2; ANOVA $F_{(2, 93)}$=4.1, P=0.02), water (L=3.36±0.7, H=1.42±0.8, U=1.28±0.3; ANOVA $F_{(2, 38)}$=3.1, P=0.05) dust bath material (L=3.77±0.7, H=0.92±0.5, U=1.31±0.4; ANOVA $F_{(2, 33)}$=7.7, P=0.00). The results show that social dominance was affected by selection for reduced fear of humans, and the welfare and reproductive ability of the birds was improved. This indicates that tameness may have played a pivotal role for the fitness and welfare of domesticated chickens.

Assessment of cattle motivation for access to pasture or feedlot environments

Caroline Lee[1], Andrew Fisher[2], Ian Colditz[1], Jim Lea[1] and Drewe Ferguson[1]
[1]CSIRO, Locked Bag 1, Armidale 2350, Australia, [2]University of Melbourne, Faculty of Veterinary Science, Victoria, 3010, Australia; caroline.lee@csiro.au

There is a public perception that cattle welfare is reduced in feedlots due to the inability to express some natural behaviours. A study was conducted to examine cattle perception of the feedlot environment by assessing motivation to access either a feedlot or a pasture environment. Two groups (n=10) of 12-month-old Angus steers were used. The Y-maze facility consisted of two mazes attached to either pasture or a feedlot. Once an animal had made a choice in the maze, it was confined to that environment (feedlot or pasture) until the next test, thus imposing a cost on its choice. Cattle underwent four stages of Y-maze testing. In stage 1, cattle spent 7 days in the feedlot and were fed a full daily ration split into two feeding times at 08:30 h and 16:00 h. Next, they were tested in the same Y-maze twice daily (at 08:30 h and 16:00 h) for 10 consecutive days. In stage 2, cattle were confined for 4 weeks in the feedlot then tested in the same Y-maze twice daily for 5 days. In stage 3, to determine the influence of lateralisation on preference, cattle were trained to learn pasture and feedlot direction in the alternate mazes and were then tested twice daily for 6 consecutive days. Stage 4 examined the effect of removing any incentive of concurrent feeding time by testing once per day at 12:00 h for 10 consecutive days and from alternating mazes. Preference for choosing the feedlot was tested by comparing the mean group observed choice to random chance (50%). The results show that when a feed reward coincided with the time of Y-maze testing, cattle showed a preference for the feedlot environment (stages 1, 2 and 3). The time of day did not influence preference. In stage 4, when no feed reward was offered in the feedlot, cattle showed a preference for the feedlot on days 2, 3 and 4 and no preference observed for the remaining 6 days. Cattle showed a preference for the feedlot when testing coincided with feeding despite being subsequently unable to access the pasture environment until the next test. Cattle did not show a preference for the pasture environment throughout the testing period. These findings should be considered in the context of the testing model which utilised a Y-maze with a feed reward at the time of testing. When the feed reward did not coincide with testing, preference for the feedlot was reduced. This suggests that the feed reward was influencing the decision to enter the feedlot. We conclude that cattle have a stronger motivation to enter a feedlot environment offering a feedlot ration than a paddock environment offering pasture.

Social housing reduces fear of novelty and improves reversal learning performance in dairy calves

Rebecca K. Meagher, Rolnei R. Daros, Joao H.C. Costa, Marina A.G. Von Keyserlingk and Daniel M. Weary

University of British Columbia, Animal Welfare Program, Faculty of Land & Food Systems, 2357 Main Mall, Vancouver, BC V6T 1Z4, Canada; rkmeagher@gmail.com

Housing unweaned calves individually is the norm on many dairy farms. However, individual housing has been shown to increase symptoms of anxiety in open field tests and, more recently, to impair performance in certain cognitive tasks compared to pair or group-housed calves. We investigated the relationship between these two effects, and whether social contact at 6 weeks old is too late to avoid these effects. Bull calves were housed individually (n=10), in pairs in double pens from 3-5 days old (Early Pair; n=12), individually until 42 days and then paired (Late Pair; n=12), and in groups of 4 or more calves with their mothers (Group; n=10; suckling was prevented using udder nets). Calves were trained to approach screens of one colour (red/white) to receive milk and avoid screens of the other colour; at c. 49 days of age, these colours were reversed. The number of sessions until they completed two consecutive perfect sessions was recorded up to a maximum of 22 sessions. Ten-minute novel object tests were conducted at c. 41 days old, before Late Pair calves were paired, and latency to touch the object (a brightly coloured ball) was used as a measure of fearfulness. Individual calves were less likely to learn the reversal task than those housed socially from an early age, with only 20% reaching the criterion versus 76% in the Early Pair and Group treatments (Fisher's P=0.006; the latter treatments were pooled as they did not differ, P>0.2). Late Pair calves were intermediate; 58% learned the task, a value somewhat higher than Individual calves (P=0.099) but not different from the early pair and grouped calves. Group calves approached the novel object more rapidly (median and interquartile range: 123 [34.5-553] s) than both Early Pair (555 [168-600] s; P=0.03) and calves with no social contact at the time (Individual and Late Pair; 600 [223-600] s; P=0.04). Latency to touch the novel object did not relate to performance in the reversal (sessions to criterion; P>0.10). In conclusion, individual housing results in both difficulty re-learning tasks and increased anxiety, but these two effects do not appear to be causally related. Early or even late pairing during the milk feeding period may be sufficient to prevent cognitive deficits but anxiety levels of paired calves are higher than those of calves housed in a more complex social environment.

Cognitive bias in gestating sows

Kristina Horback, Megan Murray and Thomas Parsons
University of Pennsylvania, Clinical Studies, 382 West Street Road, Kennett Square, PA 19348,
USA; khorback@vet.upenn.edu

Conventional methods of measuring pig welfare often focus on overt physical ailments, such as skin lesions, lameness, and body condition that are suggestive of negative affective states. Assessments that address positive affective states of pigs are less common. A judgment bias task provides such an opportunity by attempting to measure the subjective mental state of an animal. This cognitive experiment posits that an animal will 'judge' a certain ambiguous cue as predictive of either a positive or negative outcome depending on their affective state. At our Swine Teaching and Research Center, approximately 150 gestating sows (PIC 1050, Landrace-Yorkshire cross) are housed in a dynamic group pen and fed via electronic sow feeding stations (Schauer Agrotronic Compident 7). Forty-two sows were randomly selected to be trained to a 'go/no-go' task; with location of a feed bowl within an experimental arena as the conditioned stimulus. The positive 'go' cue allowed for free access to the corn-soybean meal, while an approach of the negative 'no-go' cue elicited an unwanted punishment (i.e. red bag waved in the face). Designation of positive versus negative cue location (left or right field) was randomized for each sow. Eighteen of the 42 sows demonstrated sufficient achievement by Day 4 to be included in the cognitive bias experiment (latency to bowl for positive trials $\leq$20 s and for negative trials $\geq$70 s). In the ambiguous cue trial, the bowl was placed in the middle of the arena, equidistant to the positive and negative cues. A cognitive bias index (CBI) was calculated based on sow latency to approach the ambiguous cue such that 0 reflects a failure to approach the ambiguous cue, and a CBI of 1 represents an approach comparable to a positive cue. Mean CBI was significantly greater (P<0.01) for the 'optimistic' group (M=0.95±0.01, n=11) compared to the 'pessimistic' group (0.08±0.06, n=7). There was a significant difference in ESF feed order rank between the two groups (P=0.01), with optimistic sows having higher ranks (eating earlier in the day) (M=20.4±4.1) than pessimistic sows (M=37.5±2.7). All sows in the present study were maintained in the same loose housing environment; however, both optimistic and pessimistic cognitive biases were displayed and social status appeared to influence these biases. This work serves as a cautionary tale for future cognitive bias research that addresses influence of the environment on affective state. Our study suggests that the individual differences in coping styles may play a role assessing affective state; with certain individuals predisposed to cope better in complex environments and highlights the challenges in designing loose housing systems that maximizes the welfare of all gestating sows.

Investigating the reward cycle for play in pigs

Lena Lidfors, Negar Farhadi, Claes Anderson and Manja Zupan
Swedish University of Agricultural Sciences, Department of Animal Environment and Health,
P.O. Box 234, 532 23 Skara, Sweden; lena.lidfors@slu.se

The aim was to investigate if growing pigs showed more behaviours indicative of anticipation when trained for access to an open arena with objects than when naïve to the arena, if the arena would release play in the pigs and if they would show behaviours indicative of relaxation when they came back to their home pen. Forty undocked piglets (10 litters of Yorkshire or Yorkshire × Landrace) were from 44 days of age housed in a weaner stable with two castrated males and two females per pen (6.5 m², home pen). Within each litter two test pigs were randomly selected. After five days of acclimatization, four days of training started followed by three weeks of twice weekly testing during which the two pigs were allowed to walk to a holding pen (2.0 m²) where they were kept for three minutes. After that the mesh gate opened and the pigs walked freely in to the arena (5.8 m²) where they stayed for 15 minutes. The following objects were placed in the arena: wellingtons, brush, traffic cone, rubber pipe, ball and knotted rope. The pigs were observed (instantaneously at 30 s intervals and continuously within each 30 s). Pigs were directly after coming back to the home pen observed for 10 min. Statistical analysis was done with GLIMIX procedures testing effect of treatment week (W1: naive pigs first day of training, W2-W4: experienced pigs). In the holding pen experienced pigs performed more standing ($P<0.01$, W1:0.63, W4:0.79 mean no. recordings/min), exploring bars ($P<0.05$, W1:0.28, W4:0.38), locomotor play ($P<0.0001$, W1:0.08, W4:0.28) and social contact ($P<0.01$, W1:0.01, W4:0.15) than the naïve pigs, whereas the latter performed more moving ($P<0.01$, W1:0.35, W4:0.18). In the arena object play was the most performed behaviour, but it did not differ between experienced and naïve pigs (n.s., W1:0.76, W4:0.77). Locomotor play was higher in naïve pigs ($P<0.0001$, W1:0.30, W4:0.20), whereas social play was more common in experienced pigs ($P<0.0001$, W1:0.03, W4:0.07). The naïve pigs moved more ($P<0.0001$, W1:0.36, W4:0.13) whereas the experienced pigs explored more ($P<0.0001$, W1:0.30, W4:0.49). During the first 10 min. after the naïve vs. experienced pigs came back to their home pen they were mainly lying down (W1:0.44, W4:0.59, $P<0.0001$), followed by standing (W1:0.42, W4:0.34, $P<0.0001$), social contact with pigs left in the home pen (W1:0.16, W4:0.27, $P<0.0001$), exploring (W1:0.21, W4:0.26, n.s.), moving (W1:0.14, W4:0.04, $P<0.0001$), eating (W1:0.08, W4:0.03, $P<0.0001$) and sitting (W1:0, W4:0.04, $P<0.05$). In conclusion, experienced pigs showed several behaviours that could indicate an anticipation to enter the arena with objects where they were more active than the naïve pigs, and they were lying more when coming back to their home pen.

Laboratory rats enjoy wheel running

Sylvie Cloutier[1], Alicia K. Netter[1] and Ruth C. Newberry[2]
[1]*Washington State University, Center for the Study of Animal Well-being, Department of Integrative Physiology and Neuroscience, P.O. Box 647620, Pullman, WA, 99164-7620, USA,* [2]*Norwegian University of Life Sciences, Department of Animal and Aquacultural Sciences, P.O. Box 5003, 1432, Ås, Norway; scloutie@vetmed.wsu.edu*

We hypothesized that access to a running wheel positively impacts the affective state of laboratory rats. Pairs (n=24) of female and male Long-Evans rats were assigned to one of three conditions: no-wheel control (NO), stationary wheel (SW), or moving wheel (MW). Treatments were provided in a test arena divided into 9 squares for 10 min daily for 10 days starting at 30-35 days of age. We compared the production of 50-kHz ultrasonic vocalizations (USVs) (associated with positive affective states), time spent in the square containing the wheel or equivalent square in NO ('wheel square', as an indicator of relative attractiveness of the wheel), and activity (assessed by the number of squares crossed per min minus the time spent in the wheel when present) on days 1 and 10 of exposure to the treatments. Pairs of rats were the units of analysis. Mixed model repeated measures ANOVAs on square root transformed (50-kHz USVs) or untransformed (wheel square and activity) data were used. Production of 50-kHz USVs, time spent in the wheel square, and activity were affected by treatment ($P \leq 0.0007$), time ($P \leq 0.001$) and their interaction ($P \leq 0.01$). Production of 50-kHz USVs did not differ between treatments on day 1 but was increased by the presence of a wheel on day 10, with a moving wheel having the largest effect (mean±SE, Day 1: NO: 25±9.6 calls/10 min, SW: 68±22.4, MW: 159±64.9; Day 10: NO: 148±59.0, SW: 337±96.2, MW: 995±105.2). On day 1, both SW and MW rats spent more time in the wheel square than NO rats. Time spent in the wheel square increased from day 1 to 10 for MW but not NO or SW rats (Day 1: NO: 49±8.4 s, SW: 265±33.8, MW: 301±37.2; Day 10: NO: 75±12.3, SW: 287±23.2, MW: 438±19.0). Level of activity increased from day 1 to 10 for SW and MW but not NO rats (Day 1: NO: 11.9±0.80 squares/min, SW: 12.7±0.70, MW: 13.2±0.50; Day 10: NO: 11.7±1.37, SW: 17.2±1.36, MW: 22.5±1.78). Based on production of 50-kHz USVs, attraction to the wheel square, and general activity, our results suggest that providing a stationary running wheel has a positive impact on rats' affective state. However, it appears that a greater benefit for the welfare of laboratory rats can be obtained by providing a moving running wheel. This outcome may be related, in part, to unpredicted movements that occur when group-housed rats share the wheel. Our results have implications for environmental enrichment programs by indicating positive effects from providing physical structures that are enhanced if the structures are dynamic.

Chronic stress leads to an enhanced stress response and an increased expectation of a food reward in sheep

Else Verbeek, Drewe Ferguson, Ian Colditz and Caroline Lee
CSIRO, Animal Food and Health Sciences, Locked Bag 1, Armidale NSW 2350, Australia;
else.verbeek@csiro.au

Animals living in captivity may be exposed to different stressors. When the stressors are sustained, excessive wear and tear (allostatic load) on the system may lead to maladaptive stress responses that could have a negative impact on the affective state. We hypothesised that exposure to repeated and prolonged stressors will lead to an enhanced hypothalamic-pituitary-adrenal (HPA) axis response to a physiological challenge and a decreased expectation of a food reward when exposed to ambiguous information (pessimistic judgement bias). Sheep were trained in a spatial judgement bias task to approach a positive location cue associated with food in one corner of a 3×3 m arena and not approach a negative location cue associated with a dog in the opposite corner of the arena. Three non-trained, non-reinforced ambiguous location cues were situated between the positive and negative locations. Sheep were then divided into one of two treatments: a chronic stress treatment subjected to an uncomfortable lying surface to restrict resting behaviour for 18 h a day in individual pens or a control treatment with animals kept in a paddock in a group. Judgement bias was assessed 6 days after starting the treatments by measuring the number of animals approaching the five different locations. On day 7, animals were subjected to a corticotropin releasing hormone and arginine vasopressin (CRH/AVP) challenge. Plasma cortisol and adrenocorticotropin hormone (ACTH) concentrations were measured at baseline, 20, 40 and 60 min. Judgement bias was analysed by fitting logistic general linear models; the best model was selected by analysis of deviance and fitted different slope and intercept parameters for treatment ($\chi^2_{(2)}$=-7.371, P<0.025), mostly due to the less steep curve of the chronic stress group that approached the locations more often, suggesting an increased expectation of a food reward. Responses to the CRH/AVP challenge were analysed by a repeated measures analysis of variance. The plasma cortisol response was affected by a time × treatment interaction ($F_{(2,56)}$=9.96, P<0.001), with the chronic stress group showing a higher response at 20 min (152±8 vs 124±9 ng/ml, $F_{(1,29)}$=5.74, P<0.05), but remained elevated for a shorter period. No effects on ACTH concentrations were found. Our results showed that individual housing combined with rest deprivation leads to an increased stress response and an increased expectation of a food reward (optimistic judgement bias). The increased allostatic load in chronically stressed animals may have implications for their affective state, and should be further investigated.

Effects of prenatal stress and post-weaning enrichment on spatial learning performances in lambs

Marjorie Coulon[1], Alain Boissy[1], Raymond Nowak[2] and Frédéric Lévy[2]
[1]INRA, UMR1213 Herbivores, 63122 St-Genès Champanelle, France, [2]INRA, CNRS, Université de Tours, UMR85 PRC, 37380 Nouzilly, France; marjorie.coulon@clermont.inra.fr

The first aim of this study was to investigate the effects of repeated negative common management practices in pregnant ewes on the subsequent cognitive abilities of their lambs tested in a spatial learning task. The second aim was to explore if an enrichment of the post-weaning rearing conditions can alleviate the detrimental effects of the prenatal stress on cognitive skills. Fifty-four lambs of Romane breed were divided into 4 groups crossing two factors: lambs born from stressed ewes during pregnancy (i.e. ewes daily exposed to unpredictable and negative challenges such as social isolation, social mixing, transport, delay feeding times) or from control ewes, and lambs reared for 4 weeks after weaning under enriched housing conditions (daily announced food delivery, addition of balloon and brush) or under standard housing conditions. At the end of the post-weaning treatment lambs were exposed to a T-maze including two opposite arms ending each to a goal box containing a bucket either with or without a food reward. In a first part of the experiment, trials consisted of choice-run with the same arm being rewarded on every trial (left for half the animals and right for the other half). In a second part of the experiment, the reversal learning was tested: reward was placed in the opposite arm from the one that was rewarded during the first part of the experiment. For the 2 parts of the experiment lambs underwent one training session of 5 trials each day, 5 days a week. The criterion of success was set at least 80% of good responses per session, during 2 consecutive sessions. All lambs reached the criterion in the two parts of the experiment in 2 to 10 sessions. The number of sessions did not differ between groups. In the first part of the experiment, the lambs born from stressed ewes perform worse than the lambs born from control ewes (respectively, 3.5 ± 0.2 vs. 4.2 ± 0.2 trials over 5 per session, $F(1,38)=5.4$, $P=0.02$). In the second part of the experiment, the reversal learning induced a drop of performances in lambs born from control ewes compared to lambs born from stressed ewes (respectively, 2.9 ± 0.2 vs. 3.4 ± 0.1 trials over 5 per session, $F(1,76)=9.7$, $P<0.01$). By contrast with the prenatal conditions, the post-weaning rearing conditions had no effect on the spatial learning performances in lambs. Therefore, prenatal stress affects to some extent performances in spatial learning and working memory in lambs but these detrimental effects were not alleviated by the enriched housing conditions after weaning. These results raise questions of potential stress-induced irreversible alteration in brain functions of the lambs.

Investigating depression-like condition in horses: attentional deficits towards auditory stimuli in some riding horses

Céline Rochais[1], Séverine Henry[1], Carole Fureix[2], Cléo Beaulieu[2] and Martine Hausberger[1,3]
[1]EthoS- Ethologie Animale et Humaine, UMR 6552 CNRS Université de Rennes1, Station Biologique de Paimpont, 35380 Paimpont, France, [2]Animal Poultry Sciences department, University of Guelph, 50 Stone Road East, Guelph, ON, N1G 2W1, Canada, [3]EthoS- Ethologie Animale et Humaine, UMR 6552 CNRS Université de Rennes1, Campus de Beaulieu, 263 Avenue du General Leclerc, CS 74205, 35042 Rennes, France; celine.rochais@univ-rennes1.fr

Domestic horses often encounter chronic stress, which in humans leads to a variety of negative psychological effects, including clinical depression. A recent study described an inactive state in domestic horses, termed withdrawn hereafter and characterised by an atypical posture, states of inactivity and low responsiveness to external stimuli, which resemble the reduced engagement with the environment seen in depressed patients. To assess whether these horses are affected by a depression-like condition, we investigated whether attentional abilities, frequently impaired in depressed patients, differ between withdrawn horses (n=12; 3 mares, 9 geldings, 20-6 and 13±1 mean years old) and control non-withdrawn animals from the same stable (n=15; 2 mares, 13 geldings, 18-4 and 10±1 mean years old), all French Saddlebred. These horses were exposed once a day for 5 consecutive days to a novel auditory stimulus, broadcasted for 3 seconds in horses' home environment. We recorded standard measures of attentional state (e.g. time spent with ears, head and/or neck orientated towards the loudspeaker). Different temporal patterns of attentional responses appeared: withdrawn horses spent less time focused on the loudspeaker on the first day (Wilcoxon, $P<0.05$), while control horses similarly paid attention to stimuli over the 5 days period (Friedman $P>0.05$). Moreover, time focused on the loudspeaker was lower in withdrawn than in control horses on this first day (Mann-Whitney, $P<0.05$). Withdrawn horses therefore seem to 'switch off' from environmental stimuli compared to control horses on the first trial, which might reflect lack of attention. Altogether, with a recent study revealing anhedonia (a core symptom of clinical depression) in withdrawn horses, results suggest a 'syndrome' that resemble clinical depression in humans, and open a promising line of investigation of what altered welfare states could look like in horses.

Social dynamics of mice living in complex laboratory environments

Becca Franks[1] and James P. Curley[2]
[1]University of British Columbia, Animal Welfare Program, 2357 Main Mall, Vancouver, BC, V6T
1Z4, Canada, [2]Columbia University, Psychology, 406 Schermerhorn; 1190 Amsterdam Ave MC
5501, New York, NY 10027, USA; beccafranks@gmail.com

Enhancing the complexity of laboratory housing may improve some aspects of animal welfare (e.g. reduced stereotypies) but risk others (e.g. increased aggression). Previous research has focused on average differences in behavior in enriched versus barren enclosures, but less is known regarding individual, dyadic, and network patterns of animals living in complex, enriched laboratory environments. We sought to characterize the behavior of CD1 mice housed in same-sex, complex enclosures (n=12/group; 4 groups). Over two weeks, behavioral data on each individual in each group was collected using continuous focal animal sampling or continuous event sampling of social behavior. These data were analyzed using multilevel modeling, dyadic analyses, and social network analyses. Multilevel modeling revealed strong individual differences (behavioral stability across time) for some social behaviors (e.g. chasing; P<0.001) but other behaviors showed only marginal stability (e.g. grooming; P=0.07). Consistent with a territorial social structure in the males, we found that on a daily-level, greater space use was associated with increased aggressive encounters (P<0.001), and that overall, the most aggressive individuals also used the most space (Shannon's diversity index, P<0.001). Examination of aggressive behavior with dyadic and network analyses indicated that pairs of individuals who engaged in more huddling (resting in physical contact) were also likely to engage in less fighting and chasing (P's<0.01). These results contribute to a growing body of work revealing the value of behavior analyses that extend beyond mean-level or aggregate patterns. Exploring these dynamics in the laboratory may provide a unique opportunity to apply experimental control to various ethological questions such as the consequences of removing highly central individuals and the role of the environment in network structure. This research also highlights an added benefit of housing animals in enriched environments: access to rich and understudied aspects of animal behavior.

Emergence of cooperation in heterogeneous populations

Ramon Escobedo[1] and Annick Laruelle[2]
[1]AEPA-Euskadi, Puente de Deusto 7, 48014 Bilbao, Spain, [2]University of the Basque Country and Ikerbasque, Fundamentos del Análisis Económico I, Avenida L. Aguirre 83, 48015 Bilbao, Spain; a.laruelle@ikerbasque.org

Consciously or unconsciously, in humans, animals or microscopic organisms, cooperative behaviors emerge in almost all circumstances of life. Even if it is still difficult to understand how someone can pay a cost for another individual to receive a benefit, some light has been shed on the mechanisms of the evolution of cooperation These mechanisms often combine each other and give rise to more complex cooperative behaviors. This happens especially when the likelihood of type (e.g. phenotype) recognition abilities is taken into account, where it is plausible to think that individuals tend to cooperate more frequently with those of the same type than their own, and defect with those of the other type. Individuals are not always capable of recognizing their own kin, or even their own species, although they can perceive a phenotypical difference between neighbor individuals. In an experiment in animal welfare in poultry industry carried out by Dennis *et al.* with domestic fowls, two types of individuals are generated by artificially marking a given proportion of fowls on the back of their necks. Chickens behave under partial information conditions: they are unable to identify their own type (whether they have a mark or not) but observe their opponent's type. This induces an alternative form of discrimination by tags. Marked fowls suffer more aggressive events and have less body mass than their unmarked pen mates. In this paper we propose a game theory model based on the above described experiment. The particular feature of the model is that individuals fail to recognize their own type, while they are able to recognize their opponents' type. More precisely the emergence of cooperation is analyzed in 2×2 symmetric games where individuals can either cooperate or defect. We study the evolution of such populations in the framework of evolutionary game by means of the replicator dynamics. The specificity of this paper is twofold. First heterogeneous populations are considered: individuals can be classified in two groups according to their phenotypic appearance. Phenotype recognition is assumed for all individuals: individuals are able to identify the type of every other individual. Individuals are also assumed to fail to recognize their own type, and thus behave under partial information conditions. Second the replicator equations are formulated in its discrete-time form. Depending on the parameter of the game, some restriction may exist for the generation length. The stability analysis of the dynamical system is carried out and a detailed description of the behavior of trajectories starting from the interior of the state-space is provided.

Preferential associations between laying hens in nests

Nadine Ringgenberg[1], Ersnt K.F. Froehlich[2], Alexandra Harlander-Matauschek[3], Micheal, J. Toscano[1], Würbel Hanno[1] and Beatrice A. Roth[2]

[1]University of Bern, ZTHZ, Division of Animal Welfare, Burgerweg 22, 3052 Zollikofen, Switzerland, [2]FSVO, ZTHZ, Burgerweg 22, 3052 Zollikofen, Switzerland, [3]University of Guelph, Animal and Poultry Science, 50 Stone Road E., Guelph, ON, N1G 2W1, Canada; nadine.ringgenberg@vetsuisse.unibe.ch

Under commercial conditions laying hens share nests during egg laying – they are thus in close proximity to one another – and may be selective of their nest partners. To investigate if preferential associations occur in nests, we examined nest sharing in laying hens. The study consisted of 8 pens (3×4 m), each housing 30 LSL laying hens and containing a row of 6 identical group nests (0.5×0.6 m), perches, feed and water. All hens were identified with leg-mounted RFID-tags and the nests were equipped with antennas. Following 5 weeks of acclimatization, we collected data on all nest visits for the first 5 h of light for 7 days during week 24 of age. Symmetric matrices were constructed for each pen and day with both the rows and columns referring to individual hens. Each matrix entry indicated if a pair of hens was recorded in the same nest (1) or not (0) during that particular day. The structure of each daily matrix was compared to the structure of the matrix of its succeeding day using a Mantel test in R statistical software with 1000 random permutations. We found positive correlations for 96% of tests (r=0.23, P<0.01) indicating that the structure of associations was relatively stable across days (significant p-values indicate that the observed r is greater than 95% of the r's of the permuted matrices). To test the null hypothesis that individual hens associated at random with other hens a preferred/avoided associations test in SOCPROG was performed. First, simple ratio association indices were calculated for each pair of hens from the 7 matrices per pen. These indices ranged from 0 (never in the same nest) to 1 (in the same nest each day). The standard deviation (SD) of these association indices was then compared against the SD of randomized association indices generated with 30000 permutations of the daily matrices (row and column totals kept constant). The SD of association indices from the observed data of each pen was greater than 95% of those of the random data (P<0.0001) indicating that associations were not random. Pairs of hens were defined as having preferential associations if they associated for at least 4 out of 7 days (association indices ≥0.5). The mean association index of those pairs was 0.70±0.06 and involved 16.0±5.76 hens per pen. Social factors thus appear to influence nest site selection, though the identified associations may also have occurred due to the timing of laying behaviour or to similar nest preferences.

Social dominance in female beef cattle: determined by body weight or by respect for age?

Marek Špinka, Radka Šárová, Francisco Ceacero and Radim Kotrba
Institute of Animal Science, Department of Ethology, Pratelstvi 815, 104 00 Prague, Czech Republic; spinka.marek@vuzv.cz

Dominance hierarchies in groups of social animals can be based either on asymmetries that are important for agonism (such as body weight) or on more 'conventional' cues (such as age), which is respected despite it having little relationship to the animal's fighting abilities. We investigated how social dominance is determined by age and weight in a herd of 29-39 beef cows over a period of ten years, focusing on all levels of the dominance hierarchy (individual, dyadic and group). At the dyadic level, age superiority had a stronger influence on the direction of social dominance in pairs than did weight superiority. Older cows were dominant in 73.6% of those dyads studied, even when the younger cow was heavier. At the group level, the strong influence of age on dominance produced a hierarchy that was very stable and strongly transitive. At the individual level, path analysis confirmed that the dominance index of a cow was more strongly associated with her age than with her weight. Our findings show that beef cows, for the most part, do not use their physical strength to attain dominance over older, but lighter, herd mates. This results in a stable age-based hierarchy, which might serve a universally shared function that promotes the smooth functioning of the herd and/or the expression of experience by older cows.

What is an emotion? The possibilities and pitfalls of reinforcement-based definitions of emotion in animals

Elizabeth Paul and Michael Mendl

University of Bristol, Veterinary Science, Langford House, Langford, BS40 5DU, United Kingdom; e.paul@bristol.ac.uk

When William James famously asked 'What is an emotion?' he offered no detailed definition of the word itself. Instead, he simply pointed to its lay construction as a common human experience or feeling. More than a century later, psychologists and philosophers are still struggling to find an adequate definition for this ubiquitous, yet multifaceted concept. To define emotion in animals is even more challenging. Its evolution, its neural substrates, its pharmacology, its links to welfare – we need to be able to discuss, investigate and explore these issues in such a way that the human-centred link between emotions and conscious feelings is put to one side. Here, we offer operational definitions for two terms, 'positively valenced state' and 'negatively valenced state'. In doing this, we build on a number of previous attempts to define animal emotion by making use of the process of instrumental reinforcement as its foundation. Reinforcement-based definitions circumvent the two problems most commonly associated with attempts to assign, measure or otherwise discuss emotional or valenced states in animals: (1) the equating of animal emotions with feeling states that can only be reported by humans; (2) the labelling of animals' behaviour patterns or physiological responses as 'emotional' simply because they resemble those seen in humans when emotional feelings are reported. Instead, reinforcement-based definitions seat emotions solidly in the domain of the events (stimuli) that can elicit them. Such definitions do not, however, simply label behaviours or physiological responses as emotional because they occur following events that we expect to be either fitness-enhancing or fitness-reducing. They leave the animal itself to demonstrate, by the work that it is willing to do, which events can be said to be able to elicit positively valenced or negatively valenced states. While advocating the use of such reinforcement-based definitions, we also draw attention to some problems associated with these approaches, and explain how definitions of valenced states can be accommodated alongside contemporary conceptualisations of 'value' in decision-making.

Anxious animals? A potential method for assessing awareness of emotions in nonhuman animals

Walter Sánchez-Suárez[1], Liz Paul[2], Melissa Bateson[3] and Georgia Mason[1]
[1]University of Guelph, 50 Stone Road, N1G 2W1 Guelph, Canada, [2]University of Bristol, Langford House, BS40 5DU Bristol, United Kingdom, [3]University of Newcastle, Framlington Place, NE2 4HH Newcastle, United Kingdom; wsanchez@uoguelph.ca

Consciously experienced emotions are feelings with 'valence' (positive/preferred or negative/aversive). Many assume that adult mammals and birds have them; however, this is far from clear scientifically and in immature animals/other taxa it is even harder to judge who has conscious emotions (cf. debates on whether fish, octopodes and crustacea feel pain). This issue is key for identifying which animals deserve moral consideration. Here we describe work aimed at identifying a behavioural task that may require conscious emotions, and so could be used to identify beings likely to have them. Our approach was inspired by studies published >20 years ago showing that trained rats and pigs can: (1) utilise drug-modified emotional states (e.g. anxiety; pain) as discriminative stimuli (DSs) in operant paradigms to determine which of two specific operants should be performed to obtain a food reward; and (2) generalize from these drug DSs with emotional components to apparently emotional non-drug-induced states (e.g. from states induced by anxiogenic drugs to those induced by electric shock or aggressive defeat). When animals use drug- or experience-induced emotions as DSs, they have been hypothesised to be consciously aware of them. To test this hypothesis, we mined data from 41 experiments on humans subjected to similar drug discrimination tasks and also asked to self-report their feelings. We regressed the lowest dose that could be used as a DS (dept. var.) against the lowest dose causing reportable feelings (indept. var.). We found a tight correlation between these two types of threshold dose ($F1,36=602.48$, $P<0.0001$). Furthermore, the slope did not differ from one, and the two thresholds did not differ ($F1,38=0.307,P=0.583$). So far this suggests that the ability to use internal states as DSs is indeed a 'Type-C' process in humans: one dependent on conscious awareness. Currently we are analysing individual-level data from 50 human subjects involved in four drug trials, to see if they yield the same pattern. If they do, this would show that humans' abilities to use internal feelings as DSs involve awareness of those feelings. Given the behavioural, neuroanatomical and neurophysiological homologies between humans and other mammals, the same may also hold for rats and pigs using their emotional states in similar tasks. This operant-based approach could therefore potentially become a tool for both asking animals what they consciously like/dislike, and for identifying species that do/do not have conscious emotions.

Assessment of different methods to identify pain in horses

Diana Stucke[1], Sarah Hall[2], Beatrice Morrone[2], Mareile Große Ruse[3] and Dirk Lebelt[1]
[1]Havelland Equine Hospital, Hohenfercheserar Str. 49, 14778 Beetzsee, Germany, [2]Scotland's Rural College (SRUC), AWIN Biomarkers Lab, Roslin Institute Building, Easter Bush, EH25 9RG, United Kingdom, [3]Lund University, Centre for Mathematical Sciences, Mathematical Statistics, Box 118, 221 00 Lund, Sweden; lebelt@pferdeklinik-havelland.de

Aim of this study was to assess different ethological and physiological approaches of identifying pain in horses. Castration was chosen as a pain model because of its wide clinical relevance. 51 equine stallions undergoing castration under general anaesthesia were divided in 3 pain-relieving treatment groups: A) single perioperative administration of Flunixin (n=19), B) additional subsequent Flunixin administrations (n=21), D) like A but with additional local anaesthesia of the spermatic cord (n=11). All horses were assessed before and five times after surgery (up to 44 h) by means of a modified Composite Pain Scale (CPS), a newly developed Facial Expression Pain Scale (FEPS), faecal glucocorticoid metabolites (GCMs) and plasma cytokine profiles. The same parameters were measured in a control group C, undergoing general anaesthesia for different non-painful procedures (n=6). Data of this exploratory study was analysed with non-parametric statistical tests (Wilcoxon signed rank, Mann-Whitney-U, non-adjusted p-values). Time had an influence on CPS and FEPS with increasing scores in treatment groups A, B and D ($P \leq 0.05$) but not in group C. Scores of groups A and B were higher compared to groups D and C at 4 h (CPS: P=0.00; FEPS: P=0.00) and 8 h (CPS: $P \leq 0.02$), whereas there was no difference between groups A and B nor between groups D and C. Only in group AB there was an increase of faecal GCMs after surgery at 20 h (P=0,09). GCM increases in group AB differed from groups C (8 h: P=0.04; 20 h: P=0.06; 32 h: P=0.03) and D (32 h: P=0.09). No effects by time or treatment could be identified for the different cytokines. CPS and FEPS seem to be able to detect different levels of post castration pain differentiating from possible anaesthesia effect better than GCM and cytokines. Additional local anaesthesia of the spermatic cord was obviously associated with lower post castration pain intensity. This work is part of the EU-funded FP7-project 'Animal Welfare Indictors (AWIN)' (FP7-KBBE-2010-4).

Lambs show ear posture changes when experiencing pain

Mirjam Guesgen[1], Ngaio Beausoleil[1], Edward Minot[2], Mairi Stewart[3] and Kevin Stafford[1]
[1]*Institute of Veterinary, Animal and Biomedical Sciences, Massey University, Palmerston North 4442, New Zealand,* [2]*Institute of Agriculture and Environment, Massey University, Palmerston North 4442, New Zealand,* [3]*Innovative Farm Systems, AgResearch Limited, Ruakura Research Centre, Hamilton 3214, New Zealand; mguesgen@gmail.com*

Ears are essential for obtaining information from the environment but ear posture, or the frequency of postural changes, may also reflect various emotional states of animals. In previous studies of adult sheep, ear posture has been found to differ according to the presumed emotional valence of the situation or event. This study aimed to see whether ear postures reflected the experience of pain in lambs. The ear, and pain-related behaviour of 4-8 week old lambs (n=45) were measured before and after tail-docking using a rubber ring. Pain-related behaviour of lambs was measured for 30 minutes before and after tail-docking, whereas the ear behaviour was scored for 30 s, halfway through each recording period. As expected, all docked lambs showed an increase in the frequency of active behaviours traditionally associated with docking pain, an increase in the time spent in abnormal postures and a decrease in time spent in normal postures (Period effects all: $P<0.05$). Tail-docking was also associated with an increase in the proportion of measured time spent with Ears Backward and a decrease in the proportion of measured time spent with Ears Plane (Period effects; median ± interquartile range: Backward: pre 0.00 ± 0.11, post 0.61 ± 0.38; Plane: pre 0.61 ± 0.71, post 0.13 ± 0.31, all $P<0.0001$). There was also a significant increase in the number of changes between ear postures from pre- to post-docking (pre 5 ± 6, post 9 ± 6, $P<0.0001$). Pain is likely to be perceived as a negative, uncontrollable situation by lambs, which may explain the increase in the time spent with ears backward. Previous studies interpreted the forward ear posture as indicative of negative emotion, elicited by separation from other group members. However separation from the flock will elicit increased alertness and therefore ears forward is more likely a specific behavioural strategy used to rectify the problem of social separation. This may explain why we found no significant change in the ears forward posture after docking. This is the first study to demonstrate changes in the ear posture of lambs associated with the negative experience of pain.

The facial expression of sheep with footrot and its relationship with lameness, total lesion score and faecal cortisol

Krista Mclennan, Carlos Rebelo, Mark Holmes, Murray Corke and Fernando Constantino-Casas
University of Cambridge, Department of Veterinary Medicince, Madingley Rd, CB3 0ES Cambridge, United Kingdom; kmm55@cam.ac.uk

The lameness associated with footrot is considered a clear indicator of pain in sheep, which has substantial animal welfare and economic impacts on the industry. Being able to assess the severity and intensity of pain effectively is vital to its alleviation. The facial expressions of 58 adult sheep (mean=1.4 yrs) categorised as either treatment only group (n=20), treatment plus NSAID's group (n=20) or matched pair non-diseased control group (n=18), were evaluated using a newly developed grimace score that assessed five areas of the face (orbital tightening, ear position, cheek bulge, lip and jaw profile and nostril/ philtrum position). In addition, the lameness score, total lesion score for all four feet and a faecal sample for cortisol metabolite analysis were collected for each sheep on initial presentation (day 1) of the disease and ninety days after treatment (day 90). There was no significant difference in cortisol metabolites (ng/g) between diseased animals and their matched controls on day 1 (V=273, P=0.8754) and on day 90 (V=257, P=0.4047), and no significant correlation (rs=0.10, P=0.89) between faecal cortisol metabolites and the total facial expression score. The facial expression scale showed a good global accuracy of the pain assessment at 78% of sheep being correctly identified as either pain or no-pain, 15% of sheep were incorrectly identified as experiencing pain when they were either a control animal or were observed day 90 and only 6% of animals with footrot were missed. The average facial score from all five areas reduced significantly (t=-5.556, df=56, P<0.001) from day 1 to day 90 suggesting a reduction in pain as animals healed. As the total lesion (rs=0.19, P=0.02), and lameness scores (rs=0.22, P=0.003) increased, so did the total facial expression score suggesting a higher level of pain in sheep with a higher lameness and higher total lesion score. The results presented suggest that the sheep grimace score may be a practical tool to assess pain in sheep as it correlates well with known pain indicator; lameness when associated with footrot. The sheep grimace scale it is still in developmental stages and thus the results should be treated as such.

Pain sensitivity and lying behaviour in dairy cows with different types of hock alterations

Christoph Winckler[1], Erin Mintline[2] and Cassandra Tucker[2]
[1]University of Natural Resources and Life Sciences (BOKU), Vienna, Department of Sustainable Agricultural Systems, Gregor-Mendel-Strasse 33, 1180 Vienna, Austria, [2]University of California, Davis, Department of Animal Science, One Shields Ave, 95616 Davis, CA, USA; christoph.winckler@boku.ac.at

Alterations of the tarsal joint such as hairless spots or lesions are commonly used as measures in dairy cattle welfare assessment. Despite information regarding their prevalence, which is often high, little is known about the effect of hock alterations on the dairy cow. The aim of this study was to investigate the effect of hair loss (HL) and lesions (L, wounds/scabs) on the lateral tarsus (LT) and the lateral tuber calcis (LTC) in terms of primary and secondary hyperalgesia and resting behaviour. Cows (LT: L=8, HL=5, Control=15; LTC: L=4, HL=7, C=19) were tested at three sites in the area of interest (A: central, B: edge of alteration and, C: 3 cm dorsal to alteration). Primary hyperalgesia was measured using electronic von Frey filaments, while force applied by a mechanical actuator cuff to the contralateral leg was used to measure secondary hyperalgesia. Limb withdrawal was the endpoint of pressure application. Three readings each were taken over separate days. Effects on resting behaviour were measured using HOBO Onset G loggers. When tested at site C, cows with lesions at the lateral tarsus showed both higher primary hyperalgesia (L=11.6 g, HL=23.3 g, Control=32.2 g; P=0.048, Kruskall-Wallis, χ^2=3.9, df=1) and secondary hyperalgesia (L=3.74N, HL=3.83N, Control=5.18N; P=0.036, Kruskall-Wallis, χ^2=4.4, df=1) than control cows. No effects were found for hairless spots on the lateral tarsus. Pain sensitivity also did not differ between control cows and animals with alterations at the lateral tuber calcis. All measures of lying behaviour (e.g. total lying time, number of lying bouts and proportion of lying time spent on affected leg) did not differ. In conclusion, our results indicate an increased pain sensitivity in cows with lesions at the lateral tarsus especially in the area around the lesion while resting behaviour remained unchanged.

Do ear postures indicate positive emotional state in dairy cows?

Helen Proctor and Gemma Carder
World Society for the Protection of Animals, 222 Grays Inn Rd, WC1X 8HB, United Kingdom;
helenproctor@wspa-international.org

We aimed to identify whether ear postures were reliable indicators of a low arousal, positive emotional state in dairy cows. By conducting experimental stroking on thirteen habituated cows we induced what is suggested to be a low arousal positive state. We conducted 370 fifteen minute focals, comprised of five minutes pre-stroking, five minutes stroking and a further five minutes post-stroking. The duration of time each cow spent in four identified ear postures was recorded and the number of ear posture changes calculated for each condition. Two of the postures were alert with tense muscles, and two were relaxed with little to no muscular tension. We performed One-Way ANOVA repeated measures analyses and found that the two alert postures, EP1 and EP3, were performed for significantly longer during the pre and post-stroking conditions than during stroking (EP1; $F_{(1,2)}=241.22$, P=0.00 and EP3; $F_{(1,2)}=39.09$, P=0.00). The opposite was found for the relaxed ear postures EP2 and EP4, which were performed for significantly longer during stroking (EP2; $F_{(1,2)}=81.20$, P=0.01 and EP4; $F_{(1,2)}=169.98$, P=0.00). Furthermore, EP2 was performed for significantly longer during post-stroking than pre-stroking and EP1 for significantly less, suggesting a lasting positive effect (EP1; $F_{(1,2)}=241.22$, P=0.01 and (EP2; $F_{(1,2)}=81.20$, P=0.01). We also found a significant drop in the frequency of ear posture changes during stroking compared with the pre and post-stroking conditions ($F_{(1,2)}=17.89$, P=0.00). We conclude that types of ear postures and the number of ear posture changes could be a useful indicator of positive emotional state in dairy cows. These results indicate that relaxed ear postures and lower numbers of ear posture changes are indicative of what is suggested to be a positive, low arousal emotional state in dairy cows. These results present a tangible and quick measure which could be incorporated into on-farm welfare assessments.

Non-invasive methods based on non-linear analyses to monitor fish behavior and welfare

Harkaitz Eguiraun, Karmele Lopez De Ipina and Iciar Martinez
University of Basque Country UPV/EHU, Plentzia Marine Station, Research Center for Experimental Marine Biology and Biotechnology, Areatza s/n, 48620 Plencia, Spain; harkaitz.eguiraun@ehu.es

Fish welfare refers to its ability to enjoy freedom from hunger, discomfort, pain, injury, disease, fear and distress and to express a normal behavior. Fish welfare keeps a positive relationship with its health, growth and the quality and safety of the products. Fish behavior is known to be altered by some contaminants. Thus the aim of this work was to develop a non-invasive methodology for image acquisition, processing and nonlinear trajectory analysis of the collective fish response to a stochastic event. The method should have the potential to be applicable to monitor stressors, ie. contaminants in fish farming. The experiment was approved by The Ethical Committee for Animal Welfare No. CEBA/285/2013MG. European sea bass (Dicentrarchus labrax, 4±2 g, 8±1 cm) were provided by Grupo Tinamenor (Cantabria, Spain). Three experimental cases were used to test the methodology: C1 and C2, consisting of 81 fish each and differing only in that C2 fish were tagged with Visual Implant Elastomer and C3 with 41 tagged fish that had been treated for 2 weeks with 4 µg methylmercury/l. The fish were subjected to a 12 h/12 h dark/light photoperiod and were fed once a day INICIO Plus feed from BioMar (56% crude protein, 18% crude fat). The fish were placed in tanks (100×100×90 cm) filled up to 80.5 cm of height with 810 l of aerated seawater under direct light (2×58 W and 5,200 lm) avoiding shadows as much as possible. To record the fish´ response one camera was placed in each tank positioned exactly in the same place. The 30 s pre and 2 min 30 s post a stochastic startle response were video recorded. Noise, artifacts and occlusions were eliminated by a motion flow process. The fish group cluster´s centroid was estimated by K-means; the trajectory was calculated for all of the sequences and analyzed by Shannon entropy (dimension less) algorithm. Results showed that the Shannon entropy of the trajectory for G3 was much lower (5.363) than those of G1 (6.302) or G2 (6.286), which were very similar, thus indicating that this parameter is of potential value to discriminate groups submitted to different degrees of stress and to monitor welfare and environmental pollution. This methodology has the potential to be embedded in an on-line/real time architecture to monitor fish schools in a farm or in the wild for various applications, including environmental contaminant monitoring programs. These results correspond to three single cases, and confirmation of the usefulness of this approach requires further works with more experimental set ups and different times of exposition to the contaminants.

Infrared thermography as a measure of temperament in beef cattle: some technical constraints

Pol Llonch, Matt Turner, Malcolm Mitchell and Simon Turner
Scotland's Rural College (SRUC), Animal Behaviour and Welfare, King's Building, West Mains Road, Edinburgh, EH9 3JG, United Kingdom; pol.llonch@sruc.ac.uk

The aim of this experiment was to evaluate changes in eye (ocular) temperature using infrared thermography as a potential indicator of a stress response and related this to individual differences in temperament during handling in beef cattle. Eighty four steers (castrates) (568.9±5.73 kg body weight; Charolais or Luing), were used. Temperament was assessed three times (d10, d23 and d44) when individually confined in a crush (weigh crate). Restlessness was scored based on a scale from 1 (no response) to 6 (severe escape attempts). Upon release, two laser sensors (located 1 and 5 m from the crush opening) provided the flight speed (m/s). Changes in eye temperature before and after rumen sampling by oral intubation were assessed twice (d1 and d58). An infrared camera located 1 m from the head (90°) took a basal picture of the animal one minute before and once every minute until 6 minutes after rumen sampling. Individual variation in maximum eye temperature was obtained from the area under the curve (AUC) of the seven records. Data were analysed using PROC GLIMMIX and PROC CORR of SAS (9.3). Eye temperature increased significantly (P<0.05) after 2 min post-sampling onwards (37.3±0.05; 37.3±0.05; 37.4±0.05; 37.2±0.07; 37.7±0.07 °C, respectively) compared to baseline (36.9±0.07 °C). The average crush score (1.7±0.09) and flight speed (0.8±0.09 m/s) were significantly correlated (r=-0.36, P=0.0008). However, neither crush score nor flight speed significantly (P>0.05) correlated with the AUC of the thermal response. Changes in environmental temperature (11.0-27.0 °C) and humidity (50-90%) during assessment significantly (P<0.001) affected ocular maximum temperature. Both crush score and flight speed have been shown here and elsewhere to reliably assess cattle temperament during handling. However, individual differences in ocular thermal response did not correlate with apparent differences in temperament or the corresponding behavioural indices. A number of factors appear to influence the efficacy of data acquisition and interpretation by thermography. It is suggested that attention is paid to environmental variables in thermography studies and that their effects must be considered in subsequent analysis.

Hair cortisol as an indicator of chronic stress in growing pigs

Nicolau Casal[1], Xavier Manteca[2], Anna Bassols[3], Raquel Peña[3] and Emma Fàbrega[1]
[1]IRTA, Subprograma de Benestar Animal, Veïnat de Sies s/n, 17121 Monells, Spain, [2]UAB, Department de Ciència Animal i dels Aliments, Facultat de Veterinària, 08193 Bellaterra, Spain, [3]UAB, Departament de Bioquímica i Biologia Molecular, Facultat de Veterinària, 08193 Bellaterra, Spain; nicolau.casal@irta.cat

It has been suggested that cortisol in hair can be a useful non-invasive indicator of chronic stress in animals since cortisol is incorporated and stored inside growing hair over several days or weeks. The aim of this study was to determine whether cortisol could be detected in pig hair and could serve as an adequate chronic stress indicator. Hair samples from two regions (cranio-dorsal (D) and dorso-lumbar (L)) of 56 entire male pigs were taken at 15 (Day 1) and 22 (Day 2) weeks of age. Remixing of pigs every week was carried out and used as a chronic stress stimuli. For this purpose, three group sizes were compared (14, 21 and 28 pigs), maintaining 14 pigs always in the pen and introducing 7 or 14 pigs from another pen. Stocking density (0.99 m^2/pig) was maintained by changing the size of the experimental pen according to group size using mobile fences. In total 4 pens were used and the pigs which were considered the experimental units were those 14 always remaining in the same pen. Cortisol values ranged from 6.4-43.88 pg/mg, with a mean value of 19.30±0.63 pg/mg. An increase over time was observed (17.95±0.93 pg/mg for Day 1 and 20.50±0.85 pg/mg for Day 2, P=0.002). Cortisol levels were also significantly higher (P≤0.001) in the L region (21.06±0.83 pg/mg) than in the D region (16.93±0.92 pg/mg). Significant correlations were found between Day 1 and Day 2 in region D (r=0.442) and in region L (r=0.466). There were also significant correlations between both regions on Day 1 (r=0.595) and Day 2 (r=0.496). A positive and significant correlation was also found between L region Day 1 and D region Day 2 (r=0.497). In conclusion, this study indicates that chronic stress produced by frequent mixing could be measured by hair cortisol levels, indicating that it could be a useful non-invasive tool. However, it is important to consider potential differences between regions when sampling. Furthermore, other methodological and sampling aspects have to be considered and improved in further studies.

Assessment of fear in turkeys: test reliability and correspondence

Marisa Erasmus and Janice Swanson
Michigan State University, Dept. Animal Science, 474 S. Shaw Lane, East Lansing, MI 48824,
USA; erasmusm@msu.edu

Fearfulness is considered to be part of animal temperament or coping styles and has important implications for behaviour and welfare. However, little information is available regarding fear responses of turkeys. This study evaluated reliability of tonic immobility (TI), open field (OF), voluntary approach (VA) and novel object (NO) tests between days and between weeks; and examined correspondence among TI, OF and VA tests. Male commercial turkeys were housed in groups of four to six in 16 pens. Turkeys were tested in all tests (OF: n=60, TI: n=66, VA and NO: n=83) on two consecutive days between 4 and 6 weeks and once more between 8 and 10 weeks. TI, OF and VA responses were moderately (r_s>0.4<0.7, P<0.05) repeatable. Turkeys maintained stable fear hierarchies in OF (ambulation latency, r_s=0.66, P<0.01), TI (vocalization latency, r_s=0.51, P<0.01) and VA (peck latency, r_s=0.53, P<0.01) tests. However, NO test responses were not repeatable (r_s<0.38). Inter-test correspondence was assessed among TI, OF and VA tests using correlation and cluster analyses, by comparing turkeys' positions in a fear hierarchy across test situations, and by examining whether turkeys that were behaviourally distinct in one test also behaved differently in other tests as a measure of behavioural consistency. Spearman rank correlation coefficients between test measures were low and test measures from the same tests clustered together, indicating little inter-test correspondence. Turkeys that differed in TI responses did not differ in OF responses and vice versa, but there were some differences in VA responses between groups of turkeys that differed in OF responses. The lack of agreement among fear tests suggests that tests are measuring responses to specific stimuli associated with each test. Although TI, OF and VA tests are reliable, there is little agreement among these tests. Fear responses of turkeys therefore appear to be context-specific and multidimensional.

The transect method as an on-farm welfare assessment of commercial turkeys

Joanna Marchewka[1], Inma Estevez[1,2], Giuseppe Vezzoli[3], Valentina Ferrante[4] and Maja M. Makagon[3]
[1]Neiker-Tecnalia, Animal Production, E-01080 Vitoria-Gasteiz, Spain, [2]IKERBASQUE Basque Foundation for Science, Alameda Urquijo 36-5, 48011 Bilbao, Spain, [3]Purdue University, Animal Sciences, 125 S. Russell St. W. Lafayette, IN 47907, USA, [4]University of Milan, V. Celoria 10, 20133 Milan, Italy; jmarchewka@neiker.net

Currently, no animal-based protocol for on-farm welfare assessment of commercial turkeys is available. The birds´ size and flighty nature make obtaining a representative sample using standard methods difficult. The transect approach, recently used in broilers, provides an alternative for conducting on-farm assessments of turkeys. The method combines routine screening procedures with line transect methodology used in wildlife studies for population abundance estimation. We compared the transect approach with a traditional method (individual sampling), and load out data. Ten 19 to 20-week old hybrid turkey flocks raised under similar husbandry protocols were evaluated. Half of the flocks were housed on 'well managed' farms,, the other half were on 'sub-optimally managed' farms, as deemed by the producer based on their performance (1 flock/barn/farm). Each barn was subdivided longitudinally into four predetermined transects. Two observers walked the transects in random order and recorded the total number of immobile (I), lamed (L), aggressive towards a mate (AM) or a human (AH), with visible head (H), vent (V) or back (B) wounds, engaging in mating behaviors (M), small (SM), featherless (F), dirty (D), sick (S), agonizing (A), or dead (DE)birds per transect. The flocks were re-evaluated on the same day using individual sampling. Thirteen randomly selected birds from 2 locations per transect were scored as they took 10 steps (if able). Within 48 h we re-assessed the birds when funneled out of the barn during load out. Using ANOVAs we determined the effects of observers, method, management and their interactions on the proportions of turkeys per barn within each category. The parameters were not affected by management ($P>0.05$ for all) or observer ($P>0.05$ for most). However, a significant effect of assessment method was detected ($P<0.05$). Pairwise comparisons showed differences between individual sampling and the two other methods ($P<0.05$) for most of the parameters with the exception of AM, B, D, S, SM and V. Differences were not detected between data collected using transect walks and during load out, except DE ($P=0.0007$ and I ($P=0.007$). The higher proportion of DE and I may have been due to fatigue or flock disruption unavoidable during the load out procedure. The results suggest that transect method may be a reliable and more promising tool for on-farm turkey welfare assessment then individual sampling, since it produced results similar to load out ones, when all birds could be scored individually.

Application of a dog personality questionnaire to investigate personality development in juvenile guide dogs

Naomi Harvey, Martin Green, Gary England and Lucy Asher
The University of Nottingham, School of Veterinary Science & Medicine, Sutton Bonington, LE12 5RD, United Kingdom; svxnh@nottingham.ac.uk

Personality can be defined as individual differences in behaviour that show inter-individual consistency across time and within similar situations. The aim of this study was to investigate normative personality development in the dog using a sample of juvenile guide dogs evaluated at 5, 8 and 12 months of age. A previously validated dog personality questionnaire was completed by the dogs' training supervisors (592 dogs; 306 M: 286 F) and volunteer carers (276 dogs; 130 M: 146 F). The majority of dogs were bred by the Guide Dogs for the Blind Association UK (567 dogs). The dogs originated from 101 dams and 56 sires. Breeds included were: Golden Retriever × Labrador (n=199), Labrador (n=113), Golden Retriever (n=99), Labrador × Golden Retriever cross (n=70), other (n=111). The questionnaire comprised 56 items, scored on a visual analogue scale, categorised into 8 'traits': Excitability, General Anxiety, Animal Chase, Body Sensitivity, Distractibility, Trainability, Adaptability, and Energy Level. Cross-classified multi-level models were used to investigate the influence on the trait scores of age, sex, and breed; with sire, and litter/dam included as random effects. Sire and litter/dam showed minimal influence on the traits at all ages (<5% of variance). Age, breed and sex all had a significant impact on trait scores. Population level scores decreased with age for: Distractibility, Excitability, Animal Chase and General Anxiety (all P<0.001), whilst scores increased with age for Trainability (P<0.001) and Adaptability (P<0.05). With regards to breed, Labradors scored lowest on General Anxiety (P<0.01), whilst Golden retrievers scored highest (P<0.001). However, Golden retrievers scored lowest on Distractibility (P<0.01) and Excitability (P<0.001). Evidence was found for a decrease in trainability during adolescence (P<0.01), something that to date has not been shown in the published literature. This decrease in the trainability rating at 8 months of age was found in the questionnaire completed by the dog's carers. Interactions between sex and age (P<0.001) suggest female dogs mature slower than males in the juvenile period, with females scoring higher on Excitability, Distractibility and Animal Chase at 5 and 8 months. To the authors' knowledge, no studies exist in the published literature concerning personality development in juvenile dogs and this is the first study to document evidence of 'adolescent' behaviour in juvenile dogs. The results of this study provide new insights into the development of personality in juvenile dogs, and could have implications for applied animal behaviour within organisations where the dog is used for working purposes.

Owner perceived closeness and owner perceived intelligence of dog can predict dog's performance on object choice task

Jessica Lee Oliva[1], Jean-Loup Rault[2], Belinda Appleton[3] and Alan Lill[1]
[1]Monash University, Biological Sciences, Wellington Road, Clayton, VIC, 3800, Australia, [2]Animal Welfare Science Centre, Melbourne University, Parkville, 3052, Australia, [3]Life and Environmental Sciences, Deakin University, Waurn Ponds, 3216, Australia; jessica.oliva@monash.edu

A positive association has been found between owner-rated dog cognition and owner perceived closeness to their dog, using the Perceptions of Dog Intelligence and Cognitive Skills (PoDIaCS) survey and the Monash Dog Owner Relationship Scale (MDORS). The neuropeptide oxytocin has been associated with both non verbal intelligence and bonding and therefore, could explain this positive relationship. The aim of this study was to objectively measure social cognitive intelligence in pet dogs using the 'object choice task' on two separate testing sessions, once after intranasal oxytocin administration and once after intranasal saline administration, and correlate performance on this task with scores on three different surveys; the MDORS, the Pet Attachment Questionnaire (PAQ), which measures levels of anxious and avoidant attachment styles, and a modified version of the PoDIaCS. It was hypothesized that dogs that scored higher on the task would have owners with higher levels of self reported closeness to their dog and perceive their dogs to have greater cognitive abilities. Seventy-five (33 M; 42 F) pet dogs and their owners participated in the study over two different testing sessions, 5-15 days apart. Owners filled in the PoDIaCS before testing session 1 and the MDORS and PAQ before testing session 2. Multiple regression analysis revealed that subscale 1 of the MDORS, 'perceived emotional closeness', and the anxious subscale of the PAQ were significant predictors in 'medial distal pointing with saline' scores, $F_{15, 47}=2.99$, $P<0.01$, subscale 4, 'learned awareness of human attention', of the PoDIaCS was a significant predictor in 'gaze with oxytocin' scores, $F_{13, 47}=2.04$, $P<0.01$, and subscales 1, 'recognition of human emotions', and 6, 'contagion of human emotions', of the PoDIaCS were significant predictors in 'gaze with saline' scores, $F_{13, 48}=2.00$, $P<0.05$. Results suggests that a dogs ability to follow pointing cues and owner rated attachment are related, while a dogs ability to follow gazing cues is related to owner-rated intelligence of the dog. Findings could be useful in the selection and training of working dogs.

Using judgement bias to assess positive affect in dogs

Rebecca E. Doyle[1], Tegan Primmer[1], Rachel Casey[2], Rafael Freire[1], Sharon Nielsen[1] and Michael Mendl[2]
[1]*School of Animal and Veterinary Sciences, Charles Sturt University, Locked Bag 588, Wagga Waggga 2678, Australia,* [2]*School of Veterinary Science, University of Bristol, Langford, BS40 5DU, United Kingdom; rdoyle@csu.edu.au*

The current study aimed to investigate the potential influence of a positive handling experience on measures of judgement bias in dogs. Thirty dogs of various breed, age and both sexes were trained to differentiate between one rewarded and one unrewarded spatial location in a standard sized arena. Training was conducted over two consecutive days, and following these dogs were randomly allocated to positive or neutral treatment groups (n=15/group). The positive treatment involved interactions of 1 min each by the handler towards the dog (affectionate talking, gentle patting/scratching) before the commencement of judgement bias testing, and at regular intervals throughout the test (6 in total). Dogs in the neutral treatment group experienced a non-interacting handler for the same duration and intervals during testing. The testing itself measured the dogs' responses to three previously unseen ambiguous spatial locations between the two learnt locations. The times to approach all locations were recorded, and the test was performed again the following day with the treatment groups swapped, allowing each dog to act as their own control. The results indicated a significant treatment × location × day effect (F=3.07, P=0.01). On the first day, results were as hypothesised with positively treated dogs approaching two of the unseen locations faster than the neutral dogs (near positive: 10.5 s vs. 13.9 s, middle: 11.2 s vs. 14.4 s), indicating a more optimistic-like judgement. On the second day of testing, results were less clear, with the neutral dogs responding more optimistically to one of the locations (near positive: 13.5 s vs. 17.5 s). The results from this experiment suggest that a short-term experimentally induced positive affective states can be measured by judgement bias, but they also suggest that prior contextual learning may be influenced by affective state and affect test results.

Behavioural and physiological indicators of affective valence and arousal in companion dogs

Emma Buckland, Charlotte Burn, Holger Volk and Siobhan Abeyesinghe
Royal Veterinary College, Production and Population Health, Hawkshead Lane, Hatfield, AL9 7TA Hertfordshire, United Kingdom; elbuckland@rvc.ac.uk

Valid and reliable indicators of emotional states in animals are required for meaningful assessments of welfare and quality of life. Indicators distinguishing valence (positive-negative) and arousal (high-low) are required for accurate identification. Here, behavioural and physiological expression of dogs during contexts designed to differ in valence (positive: P, negative: N) and arousal (high: HA, low: LA). Positive contexts were obtaining food treats (PHA) and comforting stroking (PLA); negative contexts were owner separation (NHA) and muzzling (NLA). Two neutral contexts were resting (Neutral LA) and owner busy (Neutral HA). Behaviour (n=88 providing valid data) and heart rate variability (n=25) were recorded continuously for 3 minutes during context presentation; surface nasal temperature (n=110) was recorded immediately after. To identify variables which discriminated between contexts, descriptive discriminant analysis extracted a significant model with two functions explaining 73.5% of the variance (Wilk's Lambda 0.046, χ^2 1540.31, df_{190}, P<0.001). Twenty-six behaviour variables had high ($\geq$0.2) discriminant loading values, 13 of which, with numerical stability, were entered into separate generalised estimating equation models with context, breed, age and sex as predictors. Context was a significant predictor for all behaviours (P's<0.05). Vocalising was observed only in NHA and huffing or sighing was observed in both negative contexts (NHA, NLA). The mean percentage of time spent with erect ears was highest in both high arousal contexts (PHA, NHA), whereas ears pinned back was highest in both low arousal contexts (PLA, NLA). Lying behaviours differed between Neutral LA and PLA with more dorsal or lateral lying positions (and less ventral) in the latter. A higher number of tail posture and activity changes and more time spent fast and gentle tail wagging were observed in the PHA context. Effects of breed, sex and age were also found for several behaviours. Surface nasal temperature following separation was significantly lower than all other contexts (GLM: F 4.92, $df_{4,41}$, P<0.001). Mean heart rate (GLMs; Wald χ^2 53.34, df5, P<0.001), standard deviation of heart rate (Wald χ^2 45.82, df_5, P<0.001) and HV/LV ratio (Wald χ^2 38.91, DF_5, P<0.001) were significantly elevated in high arousal contexts. In conclusion, behaviour and physiology variables distinguished valence and arousal in dogs; although no indicators definitively discriminated positive valence irrespective of arousal with the experimental contexts used. In particular the ears and tail expressions best distinguished between emotional contexts.

Does dog behaviour and physiology after doing wrong indicate guilt, shame or fear?

Carla M.C. Torres-Pereira and Donald M. Broom
Centre for Animal Welfare and Anthrozoology, University of Cambridge, Department of Veterinary Medicine, Madingley Road, CB3 0ES Cambridge, United Kingdom; dmb16@cam.ac.uk

Guilt and shame can be distinguished by the extent of internal and external relevance of an action perceived to be wrong. Measurable consequences may not allow these to be distinguished but sometimes do if context is considered. Responses after wrong-doing also have to be distinguished from those caused by fear of potential retribution and efforts to elicit signs of approbation. Experimental studies, by the authors and others, of dogs in situations where they could do right or wrong, are presented as evidence for the concepts and feelings that the dogs have. Some experimental situations are described that give information about dog reactions to owners or trainers when they have done right or wrong. Other experiments described here are designed to exclude any cues from people that might change the behaviour and physiology of the dogs so as to suggest that they have a feeling of guilt. When 32 pet dogs were told not to take a readily accessible food item by an owner who then left the room, the heart-rate of the dogs that took the food was higher than that of dogs that did not (P<0.001). Dogs that took the food were more likely to look at the impassive owner after his or her return (P<0.001). In another more complex study, after receiving instructions, 32 gun-dogs had a lower heart-rate (P<0.037) if they did right. If they did wrong during a task, they were more likely to stop and look back at the trainer (P<0.001). Such evidence indicates that the dogs had a concept of whether or not they had done as instructed. It is not possible to distinguish between a feeling of guilt because of having done wrong and a feeling resulting from a prediction of potential retribution. Is such a distinction possible for humans?

Does routine human-animal-contact in beef and dairy heifers alter their reaction to a standardised veterinary examination?

Katharina L Graunke[1,2] and Jan Langbein[2]
[1]*University of Rostock, Faculty of Agricultural and Environmental Sciences (AUF), PHENOMICS office, Justus-von-Liebig-Weg 6, 18059 Rostock, Germany,* [2]*Leibniz Institute for Farm Animal Biology (FBN), Institute of Behavioural Physiology, Wilhelm-Stahl-Allee 2, 18196 Dummerstorf, Germany; graunke@fbn-dummerstorf.de*

In both beef and dairy cattle husbandry, many changes in management have reduced direct contact between humans and animals. Therefore, the habituation of the animals to their human care takers is often insufficient and causes stress and danger during necessary management procedures such as veterinary examinations, etc. The aim of this study was hence to investigate whether routine, short inspections on pasture would alter the behaviour and physiological stress reaction of beef and dairy heifers during a standardised veterinary examination. Two herds of beef heifers (each 62 Angus × Limousin × Fleckvieh crossbreeds) and two herds of dairy heifers (each 43 German Holstein) were examined 3 times in a crush by a veterinarian in a standardised procedure (VE). One beef herd and one dairy herd were inspected by the same human regularly on pasture, 32 times of each 21-30 min length spread over 8 weeks. The other two herds were controls. VE1 took place before inspections began, VE2 7 weeks into the inspection period and VE3 6 weeks after stopping the inspections. In a preliminary analysis with data of half of the heifers of each herd, we analysed the influence of inspection on differences between the VEs regarding the recorded behaviours and cortisol levels with an ANOVA (proc GLM, SAS 9.3, SAS Institute Inc., USA) with inspection yes/no as fixed factor. A Tukey-Kramer correction for multiple testing was applied. In the dairy heifers, inspection influenced ear, tail and head positions with the inspected herd showing an overall increase of relaxed ear, tail and head positions (e.g. ears relaxed: $F=5.88$, $P=0.020$; tail flapping: $F=5.30$, $P=0.027$; head position changes: $F=6.99$, $P=0.012$). These effects were also present when inspection had stopped (e.g. head position changes: $F=6.36$, $P=0.016$). In the beef heifers, inspection influenced entrance speed by keeping it at the same level (ΔVE2-VE1; $F=12.0$, $P=0.001$) while it decreased in the control. Inspection also influenced ear and head positions (e.g. ear position changes: $F=7.06$, $P=0.010$; head relaxed: $F=4.33$, $P=0.042$). After inspection was terminated, entrance time increased in the inspected herd while it stayed at the same level in the control (ΔVE3-VE2; $F=12.7$, $P=0.001$). To conclude, the short, regular inspection changed the behaviour during standardised veterinary examinations. These effects were more prevalent in the dairy heifers, where they were evident even 6 weeks after inspection stopped.

Influence of castration method and sex on lamb behaviour, the development of ewe-lamb bonding and stress responses

Katarzyna Maslowska[1,2], Eurife Abadin[3] and Cathy M Dwyer[2]
[1]Royal Dick School of Veterinary Studies, University of Edinburgh, Easter Bush Campus, EH25 9RG, Edinburgh, United Kingdom, [2]SRUC, Animal Behaviour and Welfare team, West Mains Road, EH9 3JG, Edinburgh, Scotland, United Kingdom, [3]Zoetis, 45 Poplar Rd., Parkville, Victoria 3052, Melbourne, Australia; kasia.maslowska@sruc.ac.uk

Castration alters the hormonal status of ram lambs, and most methods cause pain, which may have a longer-term impact on lamb behaviour. As part of a wider study on different castration methods, including vaccination against GnRF, we investigated whether different castration methods altered ewe-lamb bonding and response to different stressful stimuli after weaning. Seventy- two Mule (Scottish Blackface × Bluefaced Leicester) × Suffolk or Texel lambs were allocated to one of 5 groups at 2 days old (n=12 per treatment or n=24 in female group): entire males (control,C), castrated using rubber rings (RR) without anaesthesia, castrated using rubber rings and local anaesthesia (LA), immunological castration using an experimental anti-GnRF vaccine (VAC) or female group (F). Ewe-lamb bonding was recorded by one observer for 10 days in 3 time periods (1-immediately after birth; 2-at 6 weeks of age; 3-at 12 weeks of age) by scan sampling 3× a day. Duration, frequency and latency of lamb anxiety behaviours were recorded after weaning in 3 testing situations (isolation, novel object and unfamiliar human). Kruskal–Wallis tests were used to determine significant differences between treatment groups. Treatment significantly affected ewe-lamb distance in Period 1 (H=15.54, P<0.005):C lambs were further from the ewe than LA, VAC or F lambs (median distance, m[Q1-Q3]: C=3.0 [0.2-10], VAC=0.5 [0-7], LA=1 [0.1-7], RR=1.5 [0.1-8], F=1.5 [0-8]). In period 2, C lambs remained further from the ewe than F lambs (P<0.01), and tended to be further than VAC lambs (P=0.06). LA lambs were also further than F (P<0.005) and VAC (P=0.05) lambs. There were no significant differences between treatments in period 3, or in stress responses after weaning. The results suggest that castration method, and lamb sex, had some impacts on ewe-lamb bonding but there were no long term effect of early pain or testosterone exposure on lamb stress responses.

Behavioral, body postural and physiological responses of sheep to simulated sea transport motions

Eduardo Santurtun[1], Valerie Moreau[1,2] and Clive J.C. Phillips[1]
[1]University of Queensland, Centre for Animal Welfare and Ethics, School of Veterinary Science, The University of Queensland, Gatton, Queensland, 4343, Australia, [2]LaSalle Beauvais Polytechnic Institute, 19, rue Pierre Waguet, Beauvais Cedex, 30313 60026, France; esanturtun@gmail.com

Ship motion causes discomfort and stress in humans, but sheep responses are unknown despite many sheep traveling long distances by ship. We compared sheep responses to roll (side to side), heave (up and down) and pitch (front to back) movements of similar amplitude and period to a commercial livestock vessel with a control treatment using a Latin Square design with 30 minutes periods. Behavior, heart rate variability, rumination, body posture and balance were measured in four merino sheep (37±0.1 kg) restrained in pairs in a crate that was placed on a moveable platform, which generated roll and pitch motions. An electric forklift produced heave motion. Focal animals were observed continuously and a heart rate monitor was attached during each treatment. Data (means) was analyzed using a General Linear Model. Heave reduced the time sheep spent ruminating (P<0.001) and lying down (P=0.012) but increased the time they spent with their head above and under other sheep's heads (P<0.001), and standing with the back supported on the crate (P=0.006). In relation to the balance, roll caused more stepping motions than pitch and control (P<0.001). Heave and roll both increased mean heart rate and reduced inter-beat interval, compared to control (P<0.001). The inter-beat intervals in the heave treatment had a reduced RMSSD (P=0.37), high frequency band (P=0.003) and high:low frequency band ratio (P=0.010), and increased low frequency band (P=0.004). These results indicate that heave and roll require sheep to make regular behavioral and physiological adjustments to body instability. Heave also affected digestive behavior and may have reduced parasympathetic nervous system activity. A negative impact of heave and roll agrees with research on the effects of these motions on other species.

Communal rearing of rabbits affects space use and behaviour upon regrouping as adults

Stephanie Buijs, Luc Maertens and Frank A.M. Tuyttens
Institute for Agricultural and Fisheries Research, Animal Sciences Unit, Scheldeweg 68, 9090 Melle, Belgium; stephanie.buijs@ilvo.vlaanderen.be

To study the effect of early social experience on the behaviour of adult female breeding rabbits, two rearing strategies were used. Litter-reared (LITREAR) rabbits were reared with littermates and their mother only. Communally reared (COMREAR) rabbits were also reared with littermates and their mother, but from 18 days of age on they were grouped with 3 other litters and their mothers. As adults (from 3 days before their 1st parturition on) both COMREAR and LITREAR rabbits were housed in pens consisting of 4 individual units, each furnished with an elevated platform. Eighteen days post-partum the 4 units per pen were merged by removing the dividers between the units, thus creating group pens each housing 4 unfamiliar adult rabbits and their litters. Four COMREAR and 4 LITREAR groups were video recorded during the 1st 48 h after grouping, and 192 scan-samples were made per group. Fewer COMREAR than LITREAR rabbits were found within 25 cm of an adult conspecific during the 1st 24 h (32 vs. 41±2% SEM), but this difference disappeared during the 2nd 24 h (39 vs. 40±2% SEM, rearing*time interaction: $F_{1,264.1}$=4, P=0.06). The percentage of rabbits in physical contact with a conspecific was not significantly affected by rearing or time (P>0.10, mean: 12±23% STD), but COMREAR rabbits spent less time in the part of the pen to which they had been confined around parturition than LITREAR rabbits (30 vs. 39% ± 1 SEM, $F_{1,296.7}$=26, P<0.0001). Use of the elevated platform did not differ significantly between the rearing strategies (P>0.10), but was higher in the 2nd than in the 1st 24 h (26 vs. 17% ± 3 SEM, $F_{1,111.5}$=8, P<0.01). Agonistic behaviour and social sniffing were not affected by the rearing strategy (P>0.10), but were observed more often in the 1st than in the 2nd 24 h (agonistic: 1.7 vs. 0.6% ± 0.3 SEM, $F_{1,511.9}$=8, P<0.01, sniffing: 1.3 vs. 0.7% ± 0.2 SEM, $F_{1,477.7}$=4, P=0.05). COMREAR rabbits spent less time eating/drinking than LITREAR rabbits (9 vs. 12% ± 1 SEM, $F_{1,469.8}$=7, P<0.01), tended to spend less time interacting with kits (0.9 vs. 1.3% ± 0.2 SEM, $F_{1,469.7}$=3, P=0.095), and spent more time sitting/lying (73 vs. 68% ± 1 SEM, $F_{1,442.4}$=13, P<0.01). During the first 24 h only, COMREAR rabbits spent less time on cage manipulation than LITREAR rabbits (1.3 vs. 3.1% ± 0.4 SEM, $F_{1,473.5}$=4, P=0.04). In summary, communal rearing affected behaviour upon regrouping as adults, decreasing the time spent in proximity of adult conspecifics, attraction to the former home unit and activity, without significantly affecting social and contact behaviour. Whether communal rearing can aid in reducing unwanted social behaviour in group housed breeding rabbits requires further study.

Social behaviour of cattle in monoculture and silvopastoral systems in the tropics

Lucía Améndola[1], Francisco Solorio[2], Carlos González-Rebeles[1], Ricardo Améndola-Massiotti[3] and Francisco Galindo[1]
[1]*Facultad de Medicina Veterinaria y Zootecnia, Universidad Nacional Autónoma de México, Departamento de Etología, Universidad 3000, 04510 México D.F., Mexico,* [2]*Facultad de Medicina Veterinaria, Universidad Autónoma de Yucatán, Departamento de Nutrición, Mérida-Xmatkuil Km. 15.5, 97100, Mérida, Yucatán, Mexico,* [3]*Universidad Autónoma de Chapingo, km 38.5, Texcoco, 56230, Mexico; galindof@unam.mx*

Most farm animals live in groups and the social organization and interactions between individuals have an impact on their welfare. Silvopastoral systems have been implemented in the tropics as an alternative for sustainable livestock production as they can provide ecosystem services and may influence animal welfare. Therefore, the objective of this study was to compare the social behaviour of cattle in a silvopastoral system (SP) based on a more complex vegetation structure, with a high density of leucaena (*Leucaena leucocephala*), guinea grass (*Panicum maximum*), star grass (*Cynodon nlemfuensis*), and live fences (tree coverage); with that of a monoculture system (MS) based on grass (*C. nlemfuensis*), in the state Yucatán. Eight heifers (*Bos indicus × Bos taurus*) in each system were observed for four consecutive days from 7:30 h to 15:30 h, in each of three paddocks in each system, during the dry and rainy seasons. We used GLMMs to compare the frequency of affiliative and agonistic behaviours between systems and assessed differences in linearity and stability of dominance hierarchies using Landau's index and Dietz R-test, respectively. We also calculated a Directional Consistency Index (DCI) for each system/ season combination, estimating the degree of directionality in behavioural interactions. Overall, heifers in the MS interacted 38% less than heifers in the SP. Non-agonistic interactions were 62% more frequent (P=0.04) in the SP and agonistic behaviours were more frequent during the dry season (P=0.02) in the MS. Heifers in the SP had a more linear and non-random dominance hierarchy in both seasons (dry season: h'=0.964, P=0.002; rainy season: h'=0.988, P<0.001), than heifers in the MS (dry season: h'=0.571, P=0.079; rainy season: h'=0.536, P<0.310). Furthermore, the dominance hierarchy in the SP was more stable between seasons (R-test=0.779, P<0.001) than the dominance hierarchy in the MS (R-test=0.224, P=0.069). The DCI was elevated in the SP and low in the MS in both seasons. Heifers in the SP maintain more stable social hierarchies and express more positive social behaviours than heifers in the MS. These results indicate that silvopastoral systems may provide benefits on heifers' welfare giving an added value for the sustainability of the system.

What makes grizzly bears optimistic?

Heidi A. Keen[1], O. Lynne Nelson[1], Charles T. Robbins[1], Marc Evans[1], David J. Shepherdson[2] and Ruth C. Newberry[3]
[1]Washington State University, Pullman, WA, 99164, USA, [2]The Oregon Zoo, Portland, OR, 97221, USA, [3]Norwegian University of Life Sciences, Department of Animal and Aquacultural Sciences, P.O. Box 5003, 1432 Ås, Norway; ruth.newberry@nmbu.no

Cognitive bias tasks purport to assess affective states via responses to ambiguous stimuli. We hypothesized that differences in affective states of captive grizzly bears (*Ursus arctos horribilis*) produced by exposure to different environmental enrichment conditions would be revealed by differences in optimism in cognitive bias tests based on positive reinforcement. We clicker-trained six bears to respond differently (nose or paw touch) to two cues (light or dark grey cards) signalling a small or large food reward, respectively (counterbalanced across bears). Responses to ambiguous probe cues intermediate to the two trained cues were classified as optimistic, appropriate for the larger reward, or pessimistic, appropriate for the smaller reward. During 30-min enrichment sessions preceding cognitive bias testing, bears were presented with cow hide, traffic cones or no additional enrichment (control). Bears spent more time interacting with cow hide than traffic cones ($F_{1,84}=1017.1$, $P<0.001$), and engaged in less stereotyped pacing in cow hide than cone and control sessions ($F_{2,128}=34.0$, $P<0.001$). The cone and control conditions (no food value) tended to produce more optimistic responses in cognitive bias tests than the cow hide condition (with food value, $t_{7.79}=1.94$, $P=0.089$). Optimism was negatively correlated with weight gain ($r=-0.53$, $P=0.02$). Omission of food rewards reduced bear willingness to participate in testing ($F_{1,53}=7.68$, $P=0.007$), but did not dampen optimism in response to the central probe ($F_{1,49}=0.73$, $P=0.40$). Overall, bears exhibited more optimistic responses to the central probe (mean±SE, 59±20.2%) than the 50% that might occur by chance (binomial test, $P=0.03$). We conclude that optimistic response bias following interaction with different environmental enrichment items did not reflect level of satisfaction with the different enrichments. Although bears were generally optimistic, it appears that greater satiety reduced optimistic responses, suggesting that the cognitive bias tests were measuring motivation for food rewards during testing.

Risk factors for poor welfare in *Psittaciformes*: do intelligence and foraging behaviour predict vulnerable parrot species?

Heather Mcdonald Kinkaid[1], Yvonne Van Zeeland[2], Nico Schoemaker[2], Michael Kinkaid[1] and Georgia Mason[1]
[1]Animal Science Dept., University of Guelph, Guelph, ON N1G 2W1, Canada, [2]Faculty of Veterinary Medicine, Utrecht University, 3584 CL Utrecht, the Netherlands; gmason@uoguelph.ca

Some species thrive in captivity, while others are prone to poor breeding, disease and/or stereotypic behaviour. Identifying what causes these patterns yields fundamental information on ecological risk factors for welfare. Focusing on parrots, we investigated four traits suggested to predict poor welfare: high sociality, naturally long foraging times, endangeredness and ecological specialism. We also investigated intelligence, proposed as both a risk factor (increasing 'boredom'/frustration) and protective (via greater behavioural plasticity). We obtained data on feather damaging behaviour (FDB, e.g. self-plucking) and other stereotypic behaviours (SB, e.g. route-tracing) via a pet owner survey yielding information for 53 species (c.1,380 birds). Data on captive breeding came from published texts: ease of breeding (prolific/moderate/difficult) tabulated for 141 species by Schubot *et al.* (Psittacine Aviculture: Perspectives, Techniques and Research, 1992), and captive hatch rates (chicks hatched/breeding pair/p.a.) recorded for 122 species by Allen and Johnson (1990 Psittacine Captive Breeding Survey) that we corrected for natural annual fecundity. For each species, literature searches yielded data on natural peak group sizes, communal roosting, food search time, diet/habitat breadths, and reported rates of developing innovative behaviours; average brain volume corrected for body mass; and IUCN conservation status. Relationships between wild and captive variables were investigated using Mesquite software to control for phylogenetic non-independence. In parrots kept as pets or for breeding by aviculturalists, results (1-tailed unless otherwise specified) were broadly similar. Naturally long food search times predicted more FDB in pets (P=0.039) and greater breeding difficulty in aviculture systems (P=0.001). A relatively endangered IUCN status predicted lower hatch rates (P=0.042), and narrower habitat breadths predicted more difficulty breeding (P=0.012). Larger relative brain sizes predicted more SB in pets (oral: P=0.048; whole-body movements: P=0.018; both 2-tailed) and tended to predict lower hatch rates in breeders (P=0.088; 2-tailed). In breeding birds, higher innovation rates also tended to predict lower hatch rates (P=0.086; 2-tailed). Sociality did not seem to be a risk factor. Our preliminary results thus suggest that captive breeding is compromised in endangered ecological specialists, and that intelligent parrot species with naturally prolonged foraging times show increased risks of behavioural and reproductive problems in captivity.

Environmental enrichment in zoos – is the environment always enriched?

Maria Andersson[1], Anna Lundberg[1], Claes Anderson[1] and Madeleine Hjelm[2]
[1]Swedish University of Agricultural Sciences, Department of Animal Environment and Health, P.O. Bpx 234, 532 23 Skara, Sweden, [2]Ethological Zoolutions, Västergatan 9, 53372 Lundsbrunn, Sweden; maria.andersson@slu.se

Environmental enrichment (EE) should relate to animals' behavioural biology. EE has been shown to increase certain species specific behaviour as well as reducing stereotypies, however generally EE have no long-term effects. Today's theoretical framework behind implementation of EE does not allow systematic evaluation and do not produce testable predictions. We suggest that it is important to enrich zoo environments in relation to needs that are at most risk of being overlooked. We would like to propose a model that relate to the motivational state of the animals. It has been stated that there is a need for an established predictive model and we would like to suggest that Maslow's model of the behavioural needs of humans can help us predict the outcome of certain EE implementations. This model demonstrates that there are a number of basic needs to be addressed before other needs can be fulfilled. The premise in the model is that unless animal´s basic needs have been met, higher levels are of no relevance. We propose that this model is applicable to the study of EE implementation in zoo animals and predict that: (1) in conducted studies of EE, animals are not subjected to an enriched environment, but rather simply an attempt to fulfil a basic need, and (2) EE that address issues on a higher level cannot be successful until all basic needs in the base level are fulfilled. It would also be possible to test predictions suggested already by Swaisgood and Shepherdson; for example when stereotypies are most likely to develop and what form of EE could reduce certain stereotypies. To conclude, we are confident in that using the Maslow´s hierachy of needs would increase the effectiveness of implementation of EE and consequently enhance the welfare of the animals kept in zoos.

Environmentally-enriching mink increases lymphoid organ weight and differentiates between two forms of stereotypic behaviour

María Díez-León and Georgia Mason
University of Guelph, 50 Stone Rd, N1G 2W1, Canada; mdiez@uoguelph.ca

Enrichment (E) studies abound for wild zoo carnivores, but typically only measure effects on corticosteroids and stereotypic behaviour (SB), ignoring other stress-relevant variables (e.g. immune parameters). Furthermore, they are often short-term, and use E of unknown motivational significance. Using a model carnivore, American mink (Neovison vison), we raised 128 animals (64 male-female pairs) long-term in non-enriched (NE) or E cages provided with stimuli known a priori to be highly preferred. At 7 mos. all were individually-housed (adult mink are solitary) in identical standardised (S) cages (small NE cages) to allow behavioural assessment blind to treatment. After 2 wks one mink per pair was returned to its rearing cage (the other continuing to be observed in S cages for a further 2 wks), for re-observation, collection of faecal samples for corticosteroid metabolite (FCM) analysis, and additional research for another 1.5 yrs. After this they were killed, and various stress-sensitive variables measured postmortem. Two type of SB types occurred: locomotor stereotypies (LOC), e.g. pacing, shown to correlate with recurrent perseveration; and repetitive scrabbling (SCR) with forepaws, unrelated to this perseveration. As expected, E mink in their rearing conditions performed less SB than NE mink (LOC: $F_{1,60}$=30.20, P<0.001; SCR: $F_{1,60}$=18.75, P<0.001), and excreted less FCM ($F_{1,29}$=8.33, P<0.01). E mink also displayed more normal activity ($F_{1,60}$=28.33, P<0.001), partly due to E-use. Unexpectedly, E males were more inactive than NE males (t=-2.89, P<0.05). Differences also emerged between LOC and SCR SBs. In E females, E-use correlated negatively with SCR ($F_{1,14}$=4.43, P=0.05) but was unrelated to LOC; in S cages, having been E-reared protected animals from performing high levels of LOC (E- vs NE-reared: $F_{1,57}$=19.23, P<0.001), but not from performing SCR which increased such that E and NE-reared mink became indistinguishable. Finally, postmortem data showed that E mink were heavier ($F_{1,57}$=11.34, P<0.01), and had less skeletal fluctuating asymmetry ($F_{1,42}$=2.87, P<0.05) and heavier lymphoid organs important in cell-mediated immunity (thymus: $F_{1,41}$=3.43, P<0.05; spleen: $F_{1,45}$=13.11, P<0.01). None of these measures, nor FCM, covaried with LOC or SCR. Mink housed longterm with highly preferred E thus showed changes consistent with better immunity and reduced developmental stress, as well as expected reductions in SB and FCM. Individual differences in SB did not covary with other stress indicators, suggesting that within a given housing type, they reflect response styles rather than differential welfare. Finally, LOC were less sensitive than SCR to current housing and E-use, perhaps consistent with aetiology involving longterm brain changes.

Differences in neophobia between hand-reared and parent-reared barn owls (*Tyto alba*)

Luis Lezana[1,2], Rubén Hernández-Soto[2], María Díez-León[2,3], Rupert Palme[4] and David Galicia[2]
[1]Sturnus Control SL, Calahorra, 26500, La Rioja, Spain, [2]University of Navarra, Department of Environmental Biology, 31008 Pamplona, Navarra, Spain, [3]University of Guelph, Department of Animal and Poultry Science, Guelph, N1G 2W1, Canada, [4]University of Vienna, Department of Veterinary Medicine, Vienna, 1210, Austria; sturnusbird@hotmail.com

Variation in early environments is known to affect personality traits such as neophobia, which has important implications for animal welfare and conservation breeding and management. We aimed to investigate the effect of different types of rearing in the neophobic behaviour of barn owls. Barn owls are a species of interest from a conservation point of view as a model species for other endangered owls (e.g. burrowing owl); and in their own right, since they are bred in captivity for reintroduction purposes to reinforce the species (European populations are in decline) and to control rodent populations. They are also important from an educational perspective, since they are commonly displayed in falconry shows. Two groups of owls were either hand-raised (HR; n=8) or parent-raised (PR; n=8) until 4 months old, when they were moved to individual aviaries. Following two weeks of habituation, birds were exposed to three situations a priori expected to elicit different degrees of neophobia: presence of: (1) a novel object; (2) an unknown human; and (3) a predator. We recorded latency to feed and species-typical behaviours of aggression/fearfulness towards the experimental factor. We also collected faecal samples 24 h after each test for fecal corticosteroid metabolite (FCM) analyses. Because of the absence of previous validation tests on this species, several FCMs were measured. After experimentation, barn owls were group housed again. HR and PR owls did not differ in their latencies to feed in the presence of an object (W=28, P=1), but as expected PR owl took longer to feed in the presence of both human (W=10, P=0<0.05) and predator (W=5, P<0.001). Unexpectedly, HR owls' latencies to feed did not vary between object and predator exposure (W=19, P>0.05), while, as predicted, PR owls took longer to feed after exposure to a predator (W=0, P<0.001). Analyses pertaining (1) behaviours of aggression/fearfulness towards the different stimuli and (2) FCMs, are on-going and will be presented. Feeding latency results showed that rearing method influenced responses to novelty and human/predator presence in the studied species. Rearing-induced differences in neophobic responses have clear implications for the survival of released animals as well as for the welfare of animals kept in captivity for breeding purposes or public display.

Domestication related anxiety genes identified in chickens

Per Jensen, Martin Johnsson and Dominic Wright
Linköping University, IFM Biology, AVIAN Behaviour and Genomics Group, Linköping University, 58183 Linköping, Sweden; perje@ifm.liu.se

Reduced fearfulness and anxiety is one of the most prominent features in domestication, vital for the ability of animals to live and reproduce in captivity. Identifying genes contributing to this could greatly increase our understanding of how populations adapt to stressful conditions. We used a novel genetic approach, simultaneously mapping polymorphisms (i.e. genetic variants) associated with anxiety behaviour and gene expression. Open Field (OF) activity was used as an indicator of anxiety, since earlier studies have shown distinct differences in this behaviour between ancestral Red Junglefowl and domesticated chickens. Birds from an advanced intercross line (AIL) between wild and domestic birds were used to map quantitative trait loci (QTL) for a number of OF behaviours (n=572) and to map whole genome expression QTL (eQTL) in hypothalamus (n=129). Behavioural QTL and eQTL were first mapped independently and then overlaid with one another to generate initial behavioural candidate genes. Individual gene expression levels were then correlated with the behavioural trait that the eQTL overlapped with, reducing the number of candidate genes from over 200 to 10 (P<0.05, controlling for multiple testing). More than half of those had a known link with behaviour or neural development, including STK17A (a serine/threonine kinase) and GABRB2 (a GABA adrenergic receptor), both showing highly significant correlations with OF behaviour (P<0.001). Comparing our gene set with a human anxiety panel (The WTCCC Bipolar disorder dataset), we found that these two genes showed significant associations with bipolar disorder in humans. We conclude that STK17A and GABRB2 are most likely involved in domestication related reduction in anxiety in chickens.

Can peripheral nerve damage caused by tail docking lead to tail pain later in the life of pigs?

Mette S. Herskin[1], Pierpaolo Di Giminiani[2], Dale Sandercock[3], Armelle Prunier[4], Céline Tallet[4], Matt Leach[2] and Sandra Edwards[2]
[1]Aarhus University, Animal Science, Foulum, P.O. Box 50, 8830 Tjele, Denmark, [2]Newcastle University, School of Agriculture, Food & Rural Development, Newcastle upon Tyne, United Kingdom, [3]SRUC, Animal and Veterinary Science Research Group, Edinburgh, United Kingdom, [4]INRA, UMR 1348 PEGASE, 35590 Saint-Gilles, France; mettes.herskin@agrsci.dk

Tail docking in the first days of life is a common practice in the production of pigs. Tail docking is done to prevent tail biting and leads to behavioural changes indicative of pain and to later development of neuromas in the docked tail tips (Herskin et al., submitted). However, it is not known whether the early peripheral nerve damage can lead to pain later in the life of pigs. Data from adult rodents and humans suggest that neuromas may lead to spontaneous nerve activity and decreased nociceptive thresholds. Such abnormal nervous activity may play a significant role in the development and persistence of chronic pain conditions in humans. One important aspect for the possible development of pain in docked pig tails is the piglet age at the time docking. Increasing evidence suggest that experience of pain in juvenile mammals may affect their later pain processing. So far, the majority of reports have focused on short term effects of early pain experiences and suggested that, in newborns, such experience have only limited or no effect on the occurrence of later neuropathic pain. Recently, investigations in rodents have used models of neuropathic pain (e.g. spared nerve injury (SNI)), and shown that early peripheral nerve damage may lead to lower nociceptive thresholds later in life. In case of pigs, this might be seen as modified social behaviour due to the pain sensations, and might induce a general fear of humans due to the handling during docking. However, in rodents these changes have not been shown until adolescence. Due to the considerable difference in maturity at birth between rat pups, human babies (both altricial) and piglets (precocial), it is not possible to perform a direct age comparison across these species. In conclusion, tail docked pigs might experience nociceptive changes in the weeks or months after the nerve damage induced by tail docking, but more research is needed in order to clarify this. Current work within a joint European ANIWHA research project (http://farewelldock.eu/) aims to address this issue.

Tail biting in pigs: (in)consistency, blood serotonin, and responses to novelty

Winanda W. Ursinus[1,2], Cornelis G. Van Reenen[1], Inonge Reimert[2], Bas Kemp[2] and J. Elizabeth Bolhuis[2]
[1]*Wageningen UR Livestock Research, Animal behaviour and Welfare, Edelhertweg 15, 8219 PH, Lelystad, the Netherlands,* [2]*Wageningen University, Department of Animal Sciences, Adaptation Physiology Group, De Elst 1, 6708 WD, Wageningen, the Netherlands; nanda.ursinus@gmail.com*

Tail biting is usually considered a persistent maladaptive behaviour in pigs. We investigated whether the tendency to develop tail biting is related to peripheral serotonergic functioning and personality characteristics of pigs. Pigs (n=480 in five rounds) were kept in conventional farrowing pens until weaning at 4 weeks of age. Thereafter, they were housed barren (B) or straw-enriched (E). Individual pigs were exposed to a back test, a novel environment test and a novel object test in an unfamiliar environment at 2, 3.5 and 13 weeks, respectively. Blood serotonin measures were determined at 8, 9 and 22 weeks. In different phases (nursery, grower and finisher), pigs were classified as (non) tail biter based on tail biting behaviour, and as (non) victim based on tail wounds. Consistency of this classification over different phases was assessed with generalized linear mixed models. Effects of housing and associations between tail biter and victim status, blood serotonin and responses to novelty were, per phase, analysed using mixed models. Pigs were not consistently classified as tail biter over all phases post-weaning, but being a victim was (B: P<0.10-0.001; E: P<0.01). Tail biters had lower scrotonin blood platelet storage (P<0.05-0.01) and higher platelet uptake velocities (P<0.10-0.05) compared to non-tail biters, but only during the phase(s) in which they tail bit. Victims also had lower blood serotonin storage than non-victims (P<0.05-0.001). Platelet uptake velocities were higher in B pigs (P<0.05). Responses during the tests were summarized into five factors ('Early life exploration', 'Near novel object', 'Cortisol', 'Vocalizations and standing alert', and 'Back test activity') using a Principal Component Analysis. B housed tail biters consistently had lower scores of the factor 'Near novel object' (P<0.10-0.05), indicating a higher fearfulness, and tended to have a higher 'Back test activity' compared to non-tail biters (P<0.10). In conclusion, tail biting in pigs starts early in life and is difficult to predict due to its inconsistency over life whereas tail damage is more consistent. Pigs showed consistent changes in peripheral serotonin measures in phases during which they displayed tail biting, and tail biters were more fearful. Furthermore, housing may affect serotonin uptake in platelets. Further research is needed to elucidate the nature of the relationship between peripheral serotonin, fearfulness and tail biting, in order to develop successful strategies to prevent this injurious behaviour.

Factors affecting oxytocin concentration and relationship between oxytocin concentration and affiliative behaviour

Siyu Chen, Shigefumi Tanaka, Sanggun Roh and Shusuke Sato
Graduate School of Agricultural Science, Tohoku University, 232-3, Yomogita, Naruko-Onsen, Osaki, Miyagi, 9896711, Japan; chensiyu140@gmail.com

In a previous study, we identified that serum oxytocin (OT) concentrations of individual calves were different under the same rearing condition. Since OT is a positive feedback hormone, the release of OT may cause behavior normalization, which can further accelerate OT secretion. Here, we examined the effects of natural suckling and brushing by humans on the OT concentration and relationship between OT and affiliative behavior. Ten Japanese Black calves were divided into natural suckling (NS, n=5) and bucket suckling (BS, n=5) groups. NS calves were raised with dams, while BS calves were artificially raised from the age of 2 days, in single hutches. Blood samples were collected at the ages of 1 month and 1 week after weaning (2 months). Six Japanese Black cows were brushed for 3 min, for 10 days. The cows were then randomly assigned to two treatments of brushing or non-brushing. Time 0 was considered as the time brushing was completed or leaving the cow for 3 min (-3 to 0 min), with an experimenter standing beside the cow. Blood samples were collected at -6, 0, 3, 15, and 30 min. Social behavior was investigated by continuous sampling of 16 Holstein lactating cows for 3 h; the total observation time was 18 h. Blood samples were collected three times for analyzing the OT and cortisol concentrations by using enzyme-linked immunosorbent assay and EIA. One-way analysis of variance, t-test, and correlation coefficients were used for statistical analysis. OT concentrations of NS (25.5 ± 4.9 pg/ml) was significantly higher and the cortisol concentrations of NS (0.58 ± 0.25 µg/ml) was significantly lower than those of BS (16.9 ± 6.7 pg/ml and 1.00 ± 0.55 µg/ml, respectively) at 1 month ($P<0.05$). The OT concentration tended to correlate negatively with the cortisol concentration ($P=0.06$). Changes in OT concentrations with time (-6, 0, 3, 15, and 30 min) were 48.88 ± 12.4, 75.92 ± 15.8, 60.61 ± 21.6, 49.02 ± 20.4, and 51.43 ± 30.2 pg/ml, respectively, in the brushing treatment and 54.38 ± 23, 52.76 ± 7.9, 50.82 ± 21.1, 41.18 ± 6.11 and 43.37 ± 3.9 pg/ml in the non-brushing treatment. The OT concentrations at time 0 in the brushing treatment was significantly higher than those at times -6 and 0 in the non-brushing treatment ($P<0.05$ and $P<0.01$, respectively). The OT and cortisol concentrations, respectively, correlated positively with frequency of affiliative behavior ($r=0.58$, $P<0.05$) and frequency of receiving aggressive behavior ($r=0.63$, $P<0.01$).

Physiological difference between crib-biters and control horses in a standardised ACTH challenge test

Sabrina Briefer Freymond[1], Deborah Bardou[1], Elodie F. Briefer[2], Rupert Bruckmaier[3], Natalie Fouche[3], Julia Fleury[1], Anne-Laure Maigrot[1], Alessandra Ramseyer[3], Klaus Zuberbuehler[4] and Iris Bachmann[1]
[1]Agroscope – Swiss National Stud Farm, Les Longs-prés, 1580 Avenches, Switzerland, [2]Institute of agricultural sciences ETH, Zürich, Universitätstrasse, 2, 8092 Zürich, Switzerland, [3]Vetsuisse Fakultät, ISME, Langgassstrasse, 120, 3012 Bern, Switzerland, [4]Université de Neuchatel, rue Emile-Argand, 11, 2000 Neuchatel, Switzerland; sbriefer@bluewin.ch

Stereotypies are repetitive and relatively invariant pattern of behaviour, apparently functionless, which are observed in a wide range of species in captivity. They can occur when a situation exceeds the natural regulatory capacity of the organism, and particularly in situations that include unpredictability and uncontrollability (chronic stress). Many studies proposed that stereotypic behaviour may serve as coping mechanism, but results are contradictory. We measured the endocrine responsiveness of 21 crib-biters and 21 control horses in a standard ACTH challenge test, which triggers a physiological stress reaction, in order to better understand the coping process of stereotypies. Heart rate was measured continuously and saliva cortisol was taken during 3 hours every 30 minutes. We did not find any difference in heart rate or RMSSD between groups. However, crib-biters had a higher cortisol response during the ACTH challenge test (mean ± SE: crib-biters, 5.36±3.00 ng/ml; controls, 4.28±3.33 ng/ml; Linear mixed model (LMM), P<0.01). Interestingly, it seems that this difference in cortisol was largely due to the crib-biters that did not crib-bite during the test (5.9±3.06 ng/ml). Indeed, these horses had higher cortisol responses than all other horses. Our results suggest that crib-biting horses differ from control horses in their HPA axis reactivity. This difference could be a consequence of chronic stress and/or genetic predisposition. Crib-biting might be a successful coping strategy that helps horses to gain control over situations and reduce cortisol levels. We conclude that preventing them to crib-bite could be counter-productive and that it would be better to change their environment.

Vocalisation latency of neonate lambs as an indicator of vigour

Christine Morton[1], Geoffrey Hinch[1], Drewe Ferguson[2] and Alison Small[2]
[1]*University of New England, Armidale, NSW 2351, Australia,* [2]*CSIRO, Division of Animal Welfare, Locked Bag 1 Armidale, NSW 2350, Australia; cmorton5@une.edu.au*

Acoustic analysis of distress vocalisations have been widely used as a measure of neurobehavioral and vigour status in human and more recently rodent neonates. Pathological states resulting from central nervous system damage such as birth asphyxia have been associated with delayed cry latency, fundamental frequency instability and vocal dysphonation. Previously mammalian bioacoustic research has been used to evaluate caller identity, maternal-young recognition and animal effective state. This study presents data on acoustic features of neonate lamb bleats which could indicate compromised cognitive functioning and early behavioural deficits of neonate lambs in the first 12 hours post birth. In the first stage of a trial designed to investigate lamb vigour measurement, merino ewes were mated to six merino sires selected for divergent breeding values associated with lamb vigour. Application of a distress stimulus, involving isolation from the dam and release from a predator-like restraint, was applied to 169 singleton lambs at various stages up to 12 hours following birth. Assessment of early behavioural latency and test arena performance was also undertaken. Bleat response times of less than 5 seconds occurred in 57% of lambs, with male lambs more likely to have a slower bleat response than females (P<0.05). Seven percent of lambs tested did not respond with a bleat within the duration of the test (180 secs). Significant difference in bleat response time (P<0.001) was also evident between sires in tests conducted before 12 hours of age with one sire showing consistently better bleat and test arena performance over time (P<0.005). Lamb bleat latency was positively correlated with test arena return-to-ewe speed at 4 hours (r=0.28, P=0.001), 8 hours (r=0.26, P=0.002) and less at 12 hours (r=0.24, P=0.005). Bleat reaction times showed improvement over a period of 12 hours suggesting developmental maturation or possible recovery from the birth event. Bleat reaction times may reflect neonatal behavioural status associated with improved survival and have potential to be used as an indicator of lamb vigour. Bleat latency following a distress stimulus may be a practical test to indicate lamb vigour or viability within the first 12 hours of birth; however, developmental milestones and age are important factors to consider if undertaking vocal assessment of neonate lambs. Further implications of this research include development of enhanced APGAR scoring methodology for neonate ungulates, especially where time of birth is known. Other acoustic parameters, birth order and associated biochemistry are also being investigated as part of ongoing research.

How sheep in different mood states react to social stimuli varying in valence

Sabine Vögeli[1,2], Beat Wechsler[2] and Lorenz Gygax[2]
[1]Animal Behaviour, Institute of Evolutionary Biology and Environmental Studies, University of Zurich, Winterthurerstrasse 190, 8057 Zurich, Switzerland, [2]Centre for Proper Housing of Ruminants and Pigs, Federal Food Safety and Veterinary Office FSVO, Institute for Livestock Science ILS, Agroscope Tänikon, 8356 Ettenhausen, Switzerland; sabine.voegeli@agroscope.admin.ch

Long-term mood has the potential to modulate short-term emotional reactions. However, the kind of interaction is not yet sufficiently explored. We kept two groups of sheep under unpredictable, stimulus-poor or predictable, stimulus-rich housing conditions for several months to induce different mood states. To elicit short-term emotional reactions, twelve sheep from each housing group were exposed to silent video films of group mates showing agonistic interactions (presumed negative), ruminating while lying (intermediate), or feeding tolerantly from one bucket (presumed positive valence). During video presentation, we recorded frontal brain reaction by using functional near-infrared spectroscopy (fNIRS), locomotor activity and ear postures by an automated tracking system and attentiveness to the stimuli by direct observations. For the analyses we used linear-mixed models and selected the best model by means of the BIC-based model probabilities (mPr). Attentiveness of sheep towards the video films was lower the more positive the stimulus (OR=0.74), and was reduced in sheep of the predictable, stimulus-rich housing condition (OR=0.33; mPr=0.23). Frontal brain activity was reduced in sheep from the predictable, stimulus-rich group during the stimulus phase of the presumed negative valence, resulting in a peak in the change of the deoxy-haemoglobin concentrations (mPr=1.0). Irrespective of stimulus valence, sheep from the predictable, stimulus-rich group showed less locomotor activity (by a factor of 0.26; mPr=0.88), a lower proportion of backwards ears (OR=0.11; mPr=0.35), and a higher proportion of forward ears (OR=3.96; mPr=0.21). In conclusion, our results suggest that housing conditions induced differences in mood, resulting in a modulation of both attentiveness to social stimuli differing in valence, and frontal brain activity.

Food neophobia declines with social housing in dairy calves

Joao H C Costa, Rolnei R Daros, Marina A G Von Keyserlingk and Daniel M Weary
University of British Columbia, Animal Welfare Program, 2357 Main Mall, Vancouver, BC, V6T
1Z, Canada; jhcardosocosta@gmail.com

Farm animals are often reluctant to consume novel feeds. Research on laboratory rodents suggests that individual housing can increase fearfulness in animals. The aim of this study was to test the prediction that social housing reduces food neophobia in dairy calves. Holstein bull calves were either reared individually (n=18) or in a group with other calves and cows (n=18) at The University of British Columbia's Dairy Education and Research Centre. In the food neophobia tests calves were exposed to two identical buckets, one empty and the other filled with a novel food (chopped hay in Experiment 1 and chopped carrots in Experiment 2). Calves were tested 30 min/d for 3 consecutive days. Data were analysed using the MIXED procedures of SAS with repeated measures. Socially housed calves consumed more of the novel food compared with individually housed calves in both Experiment 1 and 2 (P<0.01). In Experiment 1, intake of hay averaged 34±5.4 g/d versus 17±5.3 g/d for socially versus individually housed calves and in Experiment 2, intake of chopped carrots averaged 27±5.6 g/d versus 6±6.0 g/d for the same two treatments. Social rearing decreased the latency to eat the novel feed (P<0.05). Socially housed calves began eating the hay after 1:43±1:38 versus 5:01±1:43 min:s for the individually housed calves. Latency to begin eating the chopped carrots averaged 4:24±1:03 vs 10:06±4:25 min:s for the socially versus individually housed calves. Individually raised calves tended to spend less time than socially raised calves eating the novel feed in both Experiments (2:17±1:24 vs. 3:36±1:19 min:s eating hay and 5:50±1:14 versus 2:20±0:55 min:s eating carrots; P=0.09). There was no effect of treatment on latency to approach the food bucket or the empty bucket in either experiment, and no effect of treatment on time spent sniffing the buckets. These results indicate that social housing of dairy calves reduces food neophobia. More generally, socially reared calves may better adjust to changes in feed and perhaps other aspects of their environment, relative to calves raised in individual pens and hutches.

Operant conditioning of urination by dairy calves

Alison Vaughan[1,2], Anne Marie De Passillé[2], Joseph Stookey[1] and Jeffrey Rushen[2]
[1]Western College of Veterinary Medicine, Department of Large Animal Clinical Sciences, 52 Campus Drive, University of Saskatchewan, Saskatoon, SK, S7N 5B4, Canada, [2]Agriculture and Agri-Food Canada, Pacific Agri-Food Research Centre, 6947 Highway 7, P.O. Box 1000, Agassiz, BC, V0M 1A0, Canada; alison.vaughan@usask.ca

The accumulation of feces and urine in dairy barns is a cause of cattle and human health concerns and environmental problems. It is usually assumed that cattle are not capable of controlling their defecation and urination. We tested whether calves could be taught to urinate in a location using either classical or operant conditioning. Twenty-four female Holstein calves were alternately assigned as treatment or control calves (experiment 1: n=12, median age, range = 39 d, 31-50 d; experiment 2: n=12, median age, range = 50 d, 29-64 d). Experiment 1 used classical conditioning procedures involving repeated pairing of entry into a stall and injection of a diuretic. During the training period (d 1-5) treatment calves were repeatedly placed in the stall (L: 1.5 m; W: 0.45 m; H: 1.2 m) and injected IV with diuretic (Salix, Intervet Inc. at 0.5 ml/kg BW) to induce urination. During the test period (d 6-15), calves were held in the stall for 10 min without diuretic injection, and urinations, defecations and vocalisations were recorded. The procedure was identical for control calves except saline was used in place of a diuretic. In the test period, the classically conditioned treatment calves did not urinate more than controls (P=0.41, t=0.87, paired t-test; means ± SE: 4.3±1.28 vs. 6.0±1.41 for treatment and control calves, respectively). In experiment 2, calves were trained using operant conditioning. On training days, operant calves were placed in the stall, received IV of diuretic (Salix, Intervet Inc. at 0.5 ml/kg BW) and, upon urination, were released from stall to receive approx. 250 ml milk reward. On test days, calves were placed in the stall but did not receive a diuretic; calves that urinated received the milk reward but calves failing to urinate within 15 min were given a 5 min 'time out' and received diuretic the following day. Yoked controls were never given diuretic but held in the stall for the same amount of time and received the same 'reward' or 'punishment' as their matched operant calf the previous day. Urinations, defecations and vocalisations occurring in the stall on test days were compared between treatment calves and controls. Calves trained using operant conditioning had a higher frequency of urinations in the stall than their controls (P=0.03, t=3.32, paired t-test; means ± SE=5.25±0.95 vs. 2.32±0.52 for treatment and control calves, respectively). The results of our experiment show it may be feasible to train cattle to urinate in specific areas using operant conditioning.

The effect of pair housing and enhanced milk feeding on play behaviour in dairy calves

Margit Bak Jensen[1], Linda Rosager Duve[1] and Daniel M. Weary[2]
[1]Aahus University, Department of Animal Science, Blichers Allé 20, 8830 Tjele, Denmark,
[2]University of British Columbia, Animal Welfare Program, MCML 189, 2357 Main Mall,
Vancouver, B.C.,V6T 1Z4, Canada; margitbak.jensen@agrsci.dk

Traditional management of pre-weaned dairy calves includes individual housing and restricted milk allowances; both these practices are thought to reduce the occurrence of locomotor play behaviour and individual housing prevents the occurrence of social play behaviour. The aim of the present study was to investigate the interactive effects of milk allowance and social contact on locomotor and social play. Forty-eight Holstein-Friesian calves were either individually or pair-housed in straw bedded pens of the same size (3.0×4.5m). Calves were also assigned to two milk-feeding treatments: standard (d 3-42: 5 l/d) and enhanced (d3-28: 9 l/d and d 29-42: 5 l/d). All calves were abruptly weaned at d 43. The play behaviour was recorded for 48 h beginning on d 15, 29, and 43. Variables were square root transformed before analysis, but below the back transformed values are given. The duration of locomotor play behaviour was higher in calves on the enhanced milk treatment (231, 144 and 14 s/24 h on d 15, 29 and 43, respectively) versus calves on standard milk (115, 110 and 24 s/24 h; P=0.02). The duration of locomotor play behaviour was higher among individually housed calves (118 s/24 h) versus pair-housed calves (67 s/24 h; P=0.003). Only pair-housed calves could perform social play behaviour. The duration of social play was lowest d 43 (69, 115, 12 s/24 h on d 15, 29, 43, respectively; P<0.001). The total duration of play behaviour did not differ between individually and pair-housed calves (130 (45-295) s/24 h; median and interquartile range). The results of the study show that enhanced milk feeding stimulates locomotor play behaviour and emphasises the importance of energy for calves' motivation to play. The results also suggest that pair-housed calves channel some play motivation into social rather than locomotor play behaviour.

To brush or not to brush? The use of mechanical brushes by dairy cows and its association with agonistic events

Daiana de Oliveira, Therese Rehn, Yezica Norling and Linda J. Keeling
Swedish University of Agricultural Sciences, Department of Animal Environment and Health,
P.O. Box 7068, 750 07, Uppsala, Sweden; daiana.oliveira@slu.se

It has been shown in several species that stressful experiences increase the expression of grooming behaviour. In cattle, the use of brushes has been shown to increase around the time of calving. The aim of this study was to address the use of mechanical swinging brushes in relation to agonistic interactions between cows as a potential indicator of social tension within a herd. It was predicted that cows who have just received aggression would visit the brush more often than cows who were the actors in the conflict. We observed a mixed group of 54 Swedish Red and Black Holstein cows for 5 hours per day, one day a week over an 8 week period. Frequency and time of the occurrence of agonistic interactions (pushes, horning, threats and fights) were recorded, as was the frequency, duration and time of brush use. A social index was created in order to rank cows and classify them as low ranking (LR), middle ranking (MR), and high ranking (HR) cows. Data were analysed using the GLIMMIX procedure. The majority of the cows were MR (59.3%), whereas HR and LR were 21.6% and 19.1% of the group, respectively. Among all agonistic events, 48.9% were pushes, followed by horning (28%), threats (22%) and fights (0.5%). The average duration of each brush visit was 82.05±4.7 seconds. Only social rank affected the frequency of brush use (F=5.42; P<0.0001). Within 5 minutes following an agonistic interaction, more HR cows used the brushes than MR and LR cows (means per 5 minutes ±S.E, HR=0.37±0.02; MR=0.25±0.01; LR=0.18±0.01). There were no effects of actor/ receiver or social rank on the duration of the visit. Our results do not support that brush use reflects social tension within the herd. Rather they support the view that brushes are an attractive resource in a loose housing system.

Social status and tear staining in nursery pigs

Amelia E.M. Marchant-Forde[1] and Jeremy N. Marchant-Forde[2]
[1]West Lafayette Jr/Sr High School, 1105 N. Grant Street, West Lafayette, IN 47906, USA, [2]USDA-ARS, LBRU, 125 S. Russell Street, West Lafayette, IN 47907, USA; marchant@purdue.edu

Tear staining in pigs has potential application as an on-farm welfare measure, known to be influenced by isolation and environmental enrichment, and related to measures of stress. The objective of this study was to determine if tear staining was influenced by social status. The study was approved by Purdue University Animal Care and Use Committee and carried out at Purdue's swine unit. Nursery age crossbred pigs (n=36) were assigned to 6 nursery pens at weaning, 2 weeks before this study, based on body weight and sex (3 pens of ♂, 3 pens of ♀ pigs). Pigs within a pen were from different litters. Pens measured 1.22×1.37 m giving 0.28 m^2 per pig. Pen walls were barred metal and floors were woven wire. Each pen contained one nipple drinker and one 4-space feeder containing a corn-based nursery pig diet. Both food and water were available ad libitum. The temperature was maintained at 30 °C. There was a combination of natural light and artificial lighting which was turned on between 0700 and 1530 h. The room was ventilated using 3 mechanical fans to maintain air quality. Pigs were individually marked using blue stock marker and behavior was recorded for 24 h per day over 5 days using 4 ceiling mounted infra-red video cameras (Panasonic VDI-2001C1H) connected to a computer based DVR system (IPD-DVR816) set to record in real time. At the end of behavior recording, both sides of the pigs' faces were photographed using a digital camcorder (Sony Handycam HDR-SR5) on still setting. Video data were played back and aggressive interactions were sampled using all-occurrences recording until 75 aggressive interactions in each pen had been sampled, noting down the winner and the loser in each aggressive interaction. Tear staining photographs were examined using Image-J software which enabled the calculation of the perimeter and area of staining around each eye and the averages. The aggressive interaction data were used to construct a dominance matrix. From this, the proportions of interactions won and lost were calculated for each pig in a pen. Pigs were ranked 1(highest) through 6 (lowest) based on this proportion. The relationships between social rank and area and perimeter of stain were then investigated using Spearman's Rank Correlation. Left and right eyes had similar amounts of tear staining as measured by area and perimeter. Social rank was correlated with left eye stain area (r=0.465, P=0.022), left eye stain perimeter (r=0.663 P<0.001) and average stain perimeter (r=0.507, P=0.011), but not with right eye measures. The results indicate that the amount of tear staining around the left eye is related to social status and could be a measure of the degree of social stress to which the pig is being subjected

Analysis of the phenotypic link between aggression at mixing and long-term social stability in groups of pigs

Suzanne Desire[1], Simon P. Turner[1], Richard B. D' Eath[1], Andrea Doeschl-Wilson[2], Craig R.G. Lewis[3] and Rainer Roehe[1]
[1]*SRUC, Animal and Veterinary Sciences Group, West Mains Road, Edinburgh, EH9 3JG, United Kingdom,* [2]*The Roslin Institute, Division of Genetics and Genomics, Easter Bush, Midlothian, EH25 9RG, United Kingdom,* [3]*PIC Europe, Alpha Building, London Road, Nantwich, CW5 7JW, United Kingdom; suzanne.desire@sruc.ac.uk*

Repeated mixing of growing pigs disrupts established social relationships and results in aggressive contests between group members. As aggression serves to establish dominance relationships, it is possible that increased aggression upon first mixing may facilitate the formation of social hierarchies. The objective of the study was to investigate whether there is a phenotypic link between behavioural traits of aggression at mixing and increased long-term group social stability, determined by reduced skin lesions at 3 weeks post mixing. Aggressive interactions were recorded for 24 hours after mixing, and the numbers of skin lesions (anterior, central, posterior) were obtained 24 hours (SL24h) and 3 weeks post-mixing (SL3wk) for 1,166 pigs. Between groups of animals (grouped at pen level) behavioural traits relating to the duration and number of reciprocal fights correlated positively with anterior SL24h (0.34 to 0.67; $P<0.01$) and negatively with central SL3wk (-0.28 to -0.38; $P<0.01$). For individual animals, most behavioural traits of aggressiveness (e.g. number and duration of reciprocal attacks initiated and received; proportion of fights won) correlated positively with SL24h (0.09 to 0.53; $P<0.001$), whereas the opposite association was found for SL3wk (-0.06 to -0.14; $P<0.05$). Traits associated with reciprocal aggression showed the strongest positive correlations with anterior SL24h and negative correlations with central SL3wk. The direction of the correlations suggests that groups and individuals that displayed much aggression at mixing had high SL24h and low SL3wk. Multiple linear regression models predicted that within aggressive cohorts, animals with a high fight success rate received slightly fewer SL24h than their equally aggressive, but unsuccessful pen mates, while animals that avoided aggression received the fewest SL24h. Conversely, high fight success at mixing predicted the lowest SL3wk while involvement in aggression, even when unsuccessful, predicted lower SL3wk than in animals which avoided aggression at mixing. This suggests that animals with no fighting experience are lower in social rank to those which have been defeated. These results provide evidence that increased aggression at mixing may aid stable hierarchy formation. This raises the ethical dilemma that increased acute aggression during mixing may actually decrease chronic aggression in groups and thus benefit long term welfare of the group.

Horse welfare: the use of Equine Appeasing Pheromone (EAP) during clipping to facilitate the adaptation process

Alessandro Cozzi[1], Elisa Codecasa[1], Olivier Carette[2], Jérôme Arnauld Des Lions[2], Thomas Normand[2], Florence Articlaux[1], Céline Lafont-Lecuelle[1], Cécile Bienboire-Frosini[1] and Patrick Pageat[1]
[1]IRSEA, Research Institute in Semiochemistry and Applied Ethology, Ethology and Neurosciences, Le Rieu Neuf, 84490, Saint Saturnin les Apt, France, [2]Vétérinaires des armées, Antenne vétérinaire de Fontainebleau, Rue des Archives, BP 70059, 77303 Fontainebleau, France; a.cozzi@group-irsea.com

Horses faced different stressors during their life in contact with humans. Synthetic Equine Appeasing Pheromone (EAP) facilitates the management of different fearful situations for horses. The study investigated the effect of EAP on ethological and physiological indicators in horses during a standardized clipping. 32 competition French saddle horses (3 years) were involved in the study: blinded procedure, randomized, two parallel homogeneous groups for size, sex, age (16 Placebo/ 16 EAP). The treatment was a gel applied into the nostrils 30 minutes before the clipping. Cortisol was measured before (T0), during (T1=10 min. after the beginning), at the end (T2) and after clipping (T3=15 min. after the end of the clipping in the box of horse). An Analogical Visual Scale (AVS) about the shearer perception of the risk during clipping was evaluated. Placebo (PL) and EAP groups were compared for all parameters. Results showed not significant differences for the duration of the clipping (t Student test EAP: 3,946.37 ±680.47 s; PL: 3,949.37 ±678.12 s; ddl=30; t=-0.01; P=0.99)) and for the AVS (EAP: 7.74±2.67; PL: 7.90±2.70; ddl=30; t=-0.17; P=0.86). About the horses that passed the test in the morning results showed similar value of cortisol before clipping for the two groups ((nmol/l) T0=EAP: 110.57; PL: 115.23) and a higher level in placebo group in all other steps (T1=EAP:115.32; PL:130.91; T2=EAP:97.63; PL:125.10; T3=EAP: 93.13; PL:109.68). Significant difference between the two groups about the paw bout (EAP: 3.62±4.06; PL: 0.00±0.00; ddl=17; t=-2.98; P=0.008) and the results about snort (EAP: 0.87±1.12; PL: 3.54±7.81; ddl=17; t=0.95; P=0.35) could indicate an inhibited attitude of horses in the placebo group also taking into account the cortisol results. Clipping is not insignificant, even on horses that are reported to be confident with the event. The present study highlighted the interest to use EAP to facilitate the adaptation process, without using forbidden substances.

Maternal investment and piglet survival – behavioural and other maternal traits important for piglet survival

Inger Lise Andersen
Animal and Aquacultural Sciences, P.O. Box 5003, 1432 Ås, Norway; inger-lise.andersen@nmbu.no

The aim of this presentation is to give an overview of which maternal traits are important for piglet survival, with special emphasis on behavioural traits. How well the sow takes care of her litter is influenced by many factors, such as genetic predisposition, parity, litter size, physical characteristics of the sow that affects the behaviour and mother-offspring relationship, prenatal effects, the farrowing environment, handling of sows (i.e. fear of humans) and management around the time of farrowing. Nest building activity in a crucial period before farrowing, maternal responses to piglet distress calls, and the quality of mother-young interactions shortly after birth are the maternal, behavioural traits most likely to have a direct effect on piglet survival. These behaviours differ between individuals as well as between gilts and older sows. However, the timing and methods to record this is crucial for using this successively in a breeding regime. In addition to genetic predisposition of these traits, maternal care can be stimulated through environmental factors, such as an optimal farrowing environment and opportunity to perform nest building. A high fear level towards humans affect maternal responses negatively and increases the incidence of crushing, and thus positive handling is considered a prerequisite to succeed with loose-housed sows. Movement disorders, large litters and high parities are all factors that will affect maternal behaviour negatively. With increasing litter size, sows use less time on piglet-related activities showing less maternal motivation and protectiveness. Negative side-effects of the current breeding, combined with the low success of breeding for increased piglet survival directly underpins the importance of finding ways to score maternal behavioural traits. Behavioural tests such as the piglet scream test may work well under experimental conditions, but fails to give clear results on commercial farms. On-farm methods to record behavioural traits of sows will be discussed.

Farrowing progress and cortisol level in sows housed in SWAP pens

J Hales[1], VA Moustsen[2], MBF Nielsen[2], PM Weber[1] and CF Hansen[1]
[1]University of Copenhagen, Groennegaardsvej 2, 1870 Frederiksberg, Denmark, [2]Pig Research Centre, Axeltorv 3, 1609 Copenhagen, Denmark; hales@sund.ku.dk

Housing sows in farrowing crates can cause stress and have a negative impact on the progress of farrowing, but in farrowing pens piglets may have greater risk of dying, especially in the first days of life. The objective of this study was to investigate if confinement increased saliva cortisol and had a negative impact on farrowing progress of sows in a pen, where they could be confined for a few days post farrowing (SWAP: Sow Welfare And Piglet protection). The study was conducted in a 1,200 sow piggery where sows were randomly allocated to one of the treatments: loose-loose (LL, n=48), loose-confined (LC, n=50) and confined-confined (CC, n=45). LL-sows were loose housed from placement in the farrowing unit until weaning, LC-sows were loose housed from entry to birth of the last piglet (BLP) and confined from BLP to day 4 post farrowing, and CC-sows were confined from day 114 of gestation to day 4. All sows were loose housed from day 4 to weaning. From day 114 of gestation to day 4 saliva samples were collected daily at 8 h, 13 h and 16 h. Sows were video recorded and piglet births were recorded from manual observation of video. Farrowing duration, birth interval and birth duration (time from birth of first piglet to birth of n'th piglet) was square root transformed and cortisol data were log transformed for data analyses performed using generalised linear models. Results presented are backtransformed values and 95% CI. Farrowing duration did not differ in CC (248 min (219-278)), LC (267 min (230-307)) and LL (237 min (201-277)) (P=0.54). Birth interval was also similar (CC: 16 min (14-18); LC: 17 min (15-20) and LL: 15 min (13-18), P=0.53) and so was birth duration (CC: 146 min (126-167); LC: 150 min (125-178) and LL: 129 min (105-156), P=0.47). On the day of farrowing CC had lower cortisol (19.6 nmol/l (15.8-24.2)) compared to LC (32.3 nmol/l (26.5-39.5); P=0.001) and LL (40.1 nmol/l (32.7-49.2); P<0.001). On day 1 CC again had lower cortisol (27.2 nmol/l (22.0-33.6)) than LC (36.9 nmol/l (30.2-45.1); P=0.03) and LC had lower cortisol level than LL (47.5 nmol/l (38.7-58.3); P=0.03). Day 2 post farrowing similar levels (P=0.19) were found in CC (24.1 nmol/l (19.4-29.8)) and LC (28.7 nmol/l (23.5-35.1)) whereas the level in LL (41.1 nmol/l (33.4-50.6)) remained elevated compared to CC (P<0.001) and LC (P=0.03). On day 4 cortisol levels did not differ between groups (P>0.05). In conclusion, farrowing progress did not differ between confined and loose housed sows. However, LL displayed elevated cortisol levels during the day of farrowing and the first two days post farrowing compared to LC and CC, indicating that sow physiology was affected by housing system.

Calf weaning in a partial suckling system

Julie Føske Johnsen[1], Annabelle Beaver[2], Cecilie Mejdell[1], Anne Marie De Passille[3], Jeffrey Rushen[3] and Dan Weary[4]
[1]Norwegian Veterinary Institute, Ullevålsveien 68, 0454 Oslo, Norway, [2]Cornell University, Department of Animal Science, 149 Morrison Hall, Ithaca, NY, USA, [3]University of British Columbia, Agassiz, V0M 1A0, BC, Canada, [4]University of British Columbia, Faculty of Land and Food systems, 2357 Main Mall, V6T 1Z4, Vancouver, BC, Canada; julie.johnsen@vetinst.no

Rearing a calf with the dam has health and welfare benefits, but separation and weaning are major welfare challenges. We studied how an additional source of milk vs. nutritional dependency on the dam in a partial suckling system affects calf weight gain and use of milk and grain feeders throughout weaning. Suckling only (SO, n=10) and suckling+milk feeder (SF, n=10) calves could suckle from their dam at night, but SF calves also had ad libitum access to an automated milk feeder (AMF). Feeder only (FO, n=10) calves had ad libitum access to an AMF, and were prevented from suckling the cow by an udder net. After six weeks (dam phase), cows and calves were separated (partial and total separation phase, 4 and 3 days respectively). At separation, SO calves were also provided ad libitum access to the AMF. All calves were weaned from milk gradually at the age of 7 weeks (weaning phase, one week). We hypothesized that access to an AMF during dam phase would result in less growth check at separation and weaning compared to calves that were nutritionally dependent upon the cow. Daily weight gains averaged (±SEM) 1.2±0.08 kg, 1.0±0.09 kg and 0.8±0.09 kg for FO, SF and SO calves respectively, with no effect of treatment ($F_{2, 22}$=1.64, P=0.2), phase ($F_{2, 22}$=0.72, P=0.5) or interaction between phase and treatment ($F_{4, 44}$=0.76, P=0.6). Daily milk intake during separation averaged 9.7±0.54 l, 8.3±0.83 l and 4.3±1.0 l and for FO, SF and SO respectively. Compared to the other treatments, SO calves drank less milk from the feeder: (mean difference for partial-; and total separation respectively: 8.25, P<0.001; and total separation 6.06, P=0.005, respectively). Daily grain intake during separation averaged 158±29.1 g, 245±48.5 g and 518±99.4 g for FO, SF and SO calves, respectively. A significant effect of phase ($F_{1.92, 50.1}$=63.24, P<0.001) (degrees of freedom are fractional to adjust for violations of sphericity assumptions), a close to significant phase*treatment effect ($F_{3.84, 50}$=2.2, P=0.053), and a significant effect of treatment was found ($F_{1, 26}$=3.43, P=0.047). SO calves consumed significantly more grain than did the FO group (Mean difference=195.3, P=0.03) and the SF group (Mean difference = 196.7, P=0.03). Additional access to a milk feeder during a suckling phase can provide a steady weight gain and intake profile through separation and weaning. Nutritionally dependent calves reared in a partial suckling system can compensate the loss of milk at separation with higher grain intakes.

Neonatal lamb rectal temperature, but not behaviour at handling, predict lamb survival

Cathy M Dwyer[1], Susanne Lesage[2] and Susan E Richmond[1,2]
[1]SRUC, King's Buildings, West Mains Road, Edinburgh EH9 3JG, United Kingdom, [2]University of Edinburgh, Royal (Dick) School of Veterinary Studies, Easter Bush, Roslin, EH25 9RG, United Kingdom; cathy.dwyer@sruc.ac.uk

Previously we have shown that lamb behaviour at birth, particularly time taken to stand, reach the udder and suck, are predictive of lamb survival. Further, they can be readily scored on commercial farms with indoor lambing and are heritable. These data, however, are not easy to measure in lambs born in outdoor environments where birth may not be observed, as occurs on approximately 50% of sheep farms in the UK. In this study we investigated whether behaviours recorded when lambs are weighed and tagged for management purposes might be related to lamb survival. Data were collected from 215 Scottish Blackface newborn (less than 24 hours old) lambs at handling, including ewe and lamb vocal behaviours, maternal response to lamb handling, lamb struggle scores and speed with which the lambs reunited with the ewe on release. Lamb rectal temperature, weight, sex and litter size were also recorded as was maternal parity. Mortality in the recorded lambs was 14.3% at 8 weeks of age. Ewe parity significantly affected her behaviour when her lamb was handled (P<0.001) and the speed with which the ewe and lamb were reunited after the lamb was released (P=0.011). However, there were no significant effects of any of the behavioural measures on the likelihood that a lamb would survive, and ewe parity also did not influence lamb survival in this cohort of animals. In contrast, lamb rectal temperature was significantly correlated with survival (median temperature °C [Q1-Q3]: surviving lamb=39.6 [39.2-39.9], lamb that died=39.0 [38.5-39.6], H=14.3, d.f.=1, P<0.001). Lamb rectal temperature was not correlated with any of the behavioural measures, although we have previously shown it to be related to lamb behavioural development at birth. The data suggest that measuring lamb rectal temperature may be a suitable measure to identify lambs with good survivability on farm.

Prenatal stocking densities affect fear responses and sociality in goat (*Capra hircus*) kids

Rachel Chojnacki, Judit Vas and Inger Lise Andersen
Norwegian University of Life Sciences, Department of Animal- and Aquacultural Sciences, P.O. Box 5003, 1432 Ås, Norway; rachel.chojnacki@nmbu.no

Prenatal stress (stress experienced by a pregnant mother) and its effects on offspring have been comprehensively studied but relatively little research has been done on how prenatal social stress affects farm animals such as goats. Here, we use the operational description of 'stress' as 'physical or perceived threats to homeostasis.' The aim of this study was to investigate the prenatal effects of different herd densities on the fear responses and sociality of goat kids. Pregnant Norwegian dairy goats were exposed to prenatal treatments of 1.0, 2.0 or 3.0 m^2 per animal throughout gestation. One kid per litter was subjected to two behavioral tests at the age of 5 weeks. The 'social test' was applied to assess the fear responses, sociality and social recognition skills when presented with a familiar and unfamiliar kid and the 'separation test' assessed the behavioral coping skills when isolated. The results indicate goat kids from the highest prenatal density of 1.0 m^2 were more fearful than the kids from the lower prenatal densities (i.e. made more escape attempts (separation test: P<0.001) and vocalizations (social test: P<0.001; separation test: P<0.001). This effect was more pronounced in females than males in the high density (vocalizations; social test: P<0.001; separation test: P=0.001). However, goat kids did not differentiate between a familiar and an unfamiliar kid at 5 weeks of age and sociality was not affected by the prenatal density treatment. We conclude that high animal densities during pregnancy in goats produce offspring that have a higher level of fear, particularly in females. Results are of high importance as many of the females are recruited to the breeding stock of dairy goats.

Pregnancy enhances the frequency of vocalizations in rabbit does exposed to buck

Imène Ilès

National veterinary school of Algiers, BP 161, El-Harrach, 16200, Algeria; iles_imene@yahoo.fr

The objective of this study was to determine the vocal behavior of the rabbit does when placed in contact with a male. The frequency of vocalizations was analyzed in relation to the physiological status of the female. A total of 155 nulliparous does and 35 bucks from a local population were used. The animals were reared into individual cages. The does were mated naturally and submitted to a semi-intensive reproduction rhythm, during 3 consecutive cycles. No pregnant rabbits were mated again after weaning, fixed on 35th day postpartum. The diagnosis of gestation was realized by abdominal palpation on day 12 after mating. During the vocalization-test, the female was placed into the cage of the buck, and the vocal and sexual behavior of the does was noted (vocalize or not; accept the mating or not). The test occurred on mating day and on day 12 postcoïtum. Proportional data were recorded as variable of Bernoulli (0-1). Analyze of variance was used to study the impact of the physiological status of the does (sexual receptivity, pregnancy, lactation) on their vocal behavior. The results, based on 661 observations, shows a significant correlation between vocal behavior and pregnancy (R of Spearman=0.74; P<0.001). The buck-exposed does vocalizes more frequently when they are pregnant (77.2 vs. 5.1%, for pregnant and no pregnant does respectively, P<0.0001). For all parity, pregnancy improves does vocalizations (nulliparous: +68.5%; primiparous: +86.0%; multiparous: +75.3%). In pregnant does, the reject of mating did not modify the frequency of vocalizations (73.1 vs. 74.6%, in receptive and no-receptive female, respectively, P>0.05). The lactation did not influence the frequency of vocalizations in pregnant does (85.2 vs. 86.3%, in lactating and no-lactating does respectively, P>0.05). These results suggest that the vocal behavior of the does, placed in contact with male on day 12 after mating, represents an effective indication of pregnancy.

On-farm evaluation of the copulation act in camels (*Camelus dromedaries*)

Mohamed M.A. Mohsen, S.A. Al-Shami, Marzouk A. Al-Ekna, Saad A. Al-Sultan and Ahmad A.A Alnaeen
King Faisal University, College of Veterinary Medicine and Animal Resources, Kingdom of Saudi Arabia, Box: 400 Al-Hassa 31982, Saudi Arabia; mmohamed@kfu.edu.sa

This study was carried out on a private camel farm to evaluate the copulation act, study the factors affecting copulation time, and investigate the behavioral responses of males during mating of receptive and enforced females. The obtained results indicated that the mean copulation time (min) was significantly longer (P<0.01) in case of the dominant male. The copulation time was also longer during mating of receptive females (P<0.01). The mean duration of mating time increased significantly (P<0.05) if copulation occurred from November to February than that during March to April. The male behavior during mating was almost successful if one male was found alone among the herd members. The dominant male in presence of a submissive one, succeeded to a greater extent in performing mating. The submissive male in presence of the dominant was distracted all the time and failed in performing proper intromission. The presence of the submissive male alone in the herd, improved his performance during mating. Non-receptive female camels could be deceived or enforced to the mating position. Enforcing or deceiving practices occurred among females which didn't show the signs of heat. The deceived or enforced females struggled during copulation and tried hard to end the mating situation by biting the male knee joint or the face seriously. During copulation, non-receptive females were unsteady, vocalizing loudly, moving frequently and didn't enable proper intromission easily. The male during mating of a non-receptive female, tried hard to fix her underneath by putting his foreleg over the female's forearm. At last, misdirection or wrong intromission were observed during the mating trials of non-receptive females. It is recommended to separate the male from the herd except for mating purposes, and the presence of two males at the same yard are hostile to all herd members.

Hatch time affects behavior and weight development in White Leghorn laying chickens

Pia Løtvedt and Per Jensen
Linköping University, IFM Biology, Linköping University, 581 83 Linköping, Sweden;
pialo@ifm.liu.se

In chickens, the hatch window may last 24-48 hours (up to 10% of the incubation time), and studies have shown that incubation length may affect post-hatch growth and physiology. However, little is known about effects on behavior. We therefore investigated how behavior variation correlates with hatching time in chickens. We also measured egg weight, hatch weight and post-hatch growth. 130 eggs were hatched while being video recorded for determination of individual hatching times. Hatching times did not correlate with egg weight, egg weight loss during incubation or hatch weight (Spearman rank correlations, all $P>0.05$). We chose the 20 first and 20 last hatched for further studies, and used 20 with average hatch time as controls. The chicks were weighed every week throughout the 8 week experiment. Weights were analyzed through a principal components analysis (PCA), and a negative correlation was found between hatching time and the largest PCA component, which explained 76.5% of the variance ($rs=-0.29$, $P=0.03$). In females only, this correlation was even stronger ($rs=-0.48$, $P=0.01$), whereas in males only, there was no significant correlation ($rs=-0.20$, $P=0.31$). At the age of 3-8 weeks, the birds were exposed to a number of behavioral tests, namely Open Field, Social Reinstatement, Novel Arena and Tonic Immobility. A PCA on all behavior responses yielded four components, explaining a total of 62.9% of the variance. Early hatching males scored significantly lower than middle hatchers on the second component, 'Response to novelty' (ANOVA, $P=0.03$). This was not seen in females (ANOVA, $P>0.05$). Furthermore, comparing early and middle hatching males, early hatchers tended to score lower on the first component, 'Passivity' (ANOVA, $P=0.09$). Again, no difference was observed in females. At 7 weeks of age, the animals were exposed to a spatial learning test, where they were required to walk through a maze to reach a goal (mealworms). A repeated measures ANOVA revealed a tendency for a difference in the time to reach the goal over the first four trials ($P=0.08$), with the early hatchers tending to solve the task more quickly than middle and late hatchers. This study is among the first to demonstrate a link between time of hatching and behavior in a precocial species like the chicken, and may help shed light on the evolutionary trade-offs between incubation length and post-hatch traits. The results may also be relevant from a perspective of stress coping and therefore also for animal welfare and productivity in the chicken industry. The mechanisms linking hatching time with post-hatch phenotype remain to be investigated.

Olfaction – the forgotten sense (in applied ethology)

Birte L Nielsen
INRA, UR1197 NeuroBiologie de l'Olfaction (NBO), Phase, Bât 325, 78352 Jouy en Josas, France;
birte.nielsen@jouy.inra.fr

Olfaction is the primary sensory modality for most vertebrate species kept by humans. The olfactory system has received considerable attention in neuroscience studies – and even more so in recent years as a reduced sense of smell has been found to be among the first symptoms of certain neurodegenerative disorders. In farming practice, odours are sometimes used as an aid in the management of production animals. Examples are when a lamb is coated in placental fluids from a foreign ewe to facilitate fostering, or when boar-spray is used to determine whether a sow displays oestrus behaviour to allow artificial insemination. But when it comes to applied ethology, the scientific study of olfaction is sparse. Despite many of our farm species having a highly developed sense of smell (coming from ancestors which depended largely on olfaction for survival), we do not use odours to any large extent in our study of their behaviour and welfare. This is largely due to the difficulties involved in controlling odorants and volatile compounds within an experimental protocol, but also because – among the human senses – olfaction is not one of the major modalities. I will present a brief history of olfactory behaviour research, mainly through illustrative examples of the influence of odours on learning, memory, mate choice, predator avoidance and food preferences. And I will ask the question: Are there lessons to be learned from neurobiology and the plasticity of the olfactory system? Based on our current knowledge of evolutionary biology, olfactory neuroscience, and applied behaviour science, I will put forward the argument that if our aim is to improve animal welfare as well as production, olfaction and odours should be taken into account to a much larger extent than is presently the case. By using the right odours at the right time in the housing and handling environments of farm, zoo, lab, and companion species, we may be able to create dynamic, species-specific odour-scapes. This would help us to be more successful in creating animal environments that are suitable and enriched, also in the eyes – or rather, the nose of the recipient.

Attitudes amongst veterinary students towards animal sentience: cross-sectional and longitudinal findings

Nancy Clarke[1], Liz Paul[2] and David Main[2]
[1]*World Society for the Protection of Animals, Education, 222 Gray's Inn Road, London, WC1X 8HB, United Kingdom,* [2]*University of Bristol, Animal Welfare and Behaviour Group, School of Veterinary Sciences, Langford House, Langford, Bristol BS40 5DU, United Kingdom; nancyclarke@wspa-international.org*

The veterinary student population has become predominantly female in recent years but little is known about the relationship between feminization within the veterinary context and attitudes towards animals, or how such attitudes might evolve during veterinary education. Between 2001 and 2011, two studies were conducted at a British university assessing veterinary students' beliefs about the sentient capacities of non-human animals. In Study 1, a Belief in Animal Sentience (BiAS) questionnaire was used to sample eleven consecutive cohorts (n=1045, veterinary students that enrolled between 2001 and 2011) of first-year veterinary students' beliefs about the sentience ('capacity for feeling') of ten species: Dogs, rats, bees, sheep, rabbits, lions, chickens, spiders, cats and pigs. In Study 2 the BiAS questionnaire was completed again by a subset of these students in their final years of study (n=218; veterinary students who first participated in 2004, 2006, 2007). In both Studies 1 and 2, students' beliefs in animal sentience varied according to each species' position on the phylogenetic scale and their morphological similarity to humans. In Study 1, female first-year veterinary students, relative to their male counterparts, had significantly higher sentience beliefs for all animal species, though with small effect sizes. Year of enrolment was also found to have a significant effect on veterinary students' belief in animal sentience, highlighting the need for caution when interpreting the results of cross-sectional studies. In Study 2, longitudinal findings indicated that individual veterinary students' belief in animal sentience did not change significantly with progression through veterinary education for the majority of the species included. Further research assessing veterinary students' belief in animal sentience is needed with larger and more representative populations of veterinary students from other universities within the U.K. and the relationship that this might have with welfare-relevant aspects of veterinary practice.

Effect of aviary housing characteristics on laying hen welfare and performance

Jasper L.T. Heerkens[1], Ine Kempen[2], Johan Zoons[2], Evelyne Delezie[1], T. Bas Rodenburg[3], Bart Ampe[1] and Frank A.M. Tuyttens[1,4]
[1]Institute for Agricultural and Fisheries Research (ILVO), Animal Sciences Unit, Scheldeweg 68, 9090 Melle, Belgium, [2]Provincial Centre for Applied Poultry Research, Poiel 77, 2440 Geel, Belgium, [3]Wageningen University, Behavioural Ecology Group, P.O. Box 338, 6700 AH Wageningen, the Netherlands, [4]Ghent University, Faculty of Veterinary Medicine, Salisburylaan 133, 9820 Merelbeke, Belgium; jasper.heerkens@ilvo.vlaanderen.be

In laying hen husbandry the potential of non-cage systems to be more animal-friendly than cage systems is often not fully realised on commercial farms. Laying hens kept in non-cage systems have an increased risk of sustaining multiple injuries and other welfare problems related to housing conditions. The aim of this cross-sectional study was to identify risk factors in commercial aviaries for animal based welfare measures, mortality and performance. Information on housing characteristics, management and performance in Belgian aviaries (n=47 flocks) were obtained through a questionnaire and farm records. Animal based measures (keel bone injuries, plumage condition, wounds and feet health) were measured per flock in 50 randomly selected 60-weeks old laying hens. Associations between animal based measures, performance, mortality, and possible risk factors were investigated using a linear model with a stepwise model selection procedure. The means provided are the least square means. We found a high incidence of keel fractures, wounds, lesions and feather damage, with a considerable variation between flocks. Wire-flooring (n=31) was associated with a higher incidence of keel bone fractures (0.86±0.02 vs. 0.76±0.03, P=0.005), a better feather score (14.2±0.4 vs. 12.0±0.5, P=0.001) and better wound scores on the back and vent (back: 3.70±0.06 vs. 3.43±0.08, P=0.012; vent: 3.72±0.07 vs. 3.39±0.10, P=0.009), a decreased mortality (3.3%±0.4 vs. 5.4%±0.5, P=0.003), and higher laying performance (89.0%±0.8 vs. 85.7%±1.1, P=0.021) compared with plastic-flooring (n=16). Row-type aviaries (n=37) showed a higher incidence of foot pad lesions (3.52±0.04 vs. 3.78±0.08, P=0.004) and decreased foot hyperkeratosis (1.61±0.02 vs. 1.49±0.04, P=0.021) compared with portal-type aviaries (n=10). Access to a free range had a positive effect on the feather score ($F_{1,41}$=4.619, P=0.038). These results indicate that within commercial aviaries there is considerable variation in the prevalence of various health, welfare and production problems and that this variation is related to specific housing characteristics. This suggests that further improvements to the housing design have potential to optimise the system and improve laying hen welfare and performance.

Feather pecking in laying hens during rearing and laying in non-cage systems, management practices and farmers opinions

Elizabeth Nicole (Elske) De Haas[1,2] and Bas T Rodenburg[3]
[1]INRA, University of Tours, Physiologie de la Reproduction et des Comportements, CNRS-UMR 7247, IFCE F, 37380 Nouzilly, France, [2]Wageningen University, Adaptation Physiology Group, De Elst 1, 6700AH, Wageningen, the Netherlands, [3]Wageningen University, Behavioural Ecology Group, De Elst 1, 6700AH Wageningen, the Netherlands; elske.dehaas@wur.nl

In non-cage systems, preventing feather pecking (FP) in laying hens can be challenging. We aimed to assess the risk factors of FP in relation to management practices during rearing and laying in non-cage systems. 47 rearing flocks of which 35 were followed into laying. FP behaviour (2×20 min) at five, 10 and 15 weeks of age and average feather damage (50 hens' damage to neck, back and belly) at 40 weeks of age was recorded and related to management practices. Farmers were asked at which age they think the risk of FP is highest and if they make management adjustments to prevent FP. During rearing, disruption of litter and restricted access to litter before 4-5 weeks of age increased severe feather pecking (SFP) at five weeks of age (22±4 vs. 6±3 pecks/20 min; P=0.05). High level of feather damage levels at 40 weeks of age were correlated to high SFP at five weeks of age (r=0.45; P=0.05). Housing conditions which affect SFP at rearing can, thus, create a risk for feather damage at laying. However, these effects may be counteracted when management during laying is optimized. Flocks in which specific management to prevent FP and fearfulness was applied (radio; pecking blocks) had less feather damage at 40 weeks of age compared to flocks with standard management (0.5±0.05 vs. 0.98±0.3; P<0.001). Most rearing farmers (14 out of 25) noted that FP occurred most frequently around five weeks of age. Of the laying farmers, 56% noted the importance of reducing fearfulness to prevent FP. Many farmers were aware of the risk periods, but did not adjust their management. It is important to increase famers' knowledge and supply practical solutions to prevent FP at rearing and laying.

Solitary play in offspring of laying hens with high and low feather-pecking activity

Stephanie Bourgon, Margaret Quinton and Alexandra Harlander-Matauschek
University of Guelph, Animal and Poultry Science, 50 Stone Road East, N1G2W1 Guelph, Canada; aharland@uoguelph.ca

Solitary play is likely to identify children with behavioural disorders including problems related to under-controlled, impulsive, or aggressive behavior. Included in this category is attention deficit hyperactivity disorder (ADHD). Feather pecking in laying hens has been suggested to represent an animal model for ADHD. We hypothesized that offspring of laying hens selected for high (H) feather-pecking activity perform higher rates of solitary play behaviour than offspring born to birds selected for low (L) feather-pecking activity. In groups of 10, 60 H chicks (1-day-old) and 60 L chicks were kept in identical floor pens littered with a mix of straw and wood shavings under conventional management conditions. On 2 consecutive days during wks 2-7, behavioral video observations were performed to quantify solitary play. Play behaviour was considered as any sequence that contained some or all of the following characteristics: repetition, incomplete, quick, exaggerated sequences; and that lacked final consummatory acts. Data were analyzed using PROC GLIMMIX (SAS 9.3). The number of running events – spontaneous running in circles or in straight line – was not different between the H and L chicks (22.7±2.0 vs 23.4±2.0; ns). The number of hopping events – chicks pushing off with both feet simultaneously and flapping their wings – was higher in H than in L chicks (24±1.9 vs 17±1.4; P=0.004). The total number of times an inanimate object (straw) was dropped and picked up multiple times on a given spot, and the number of times an object was dropped and picked up while the bird was running was greater in H than in L chicks (1.5±0.3 vs 0.7±0.2; P=0.033). Taken together, our results show that H birds performed higher rates of solitary play. Further research is warranted to determine whether there is an association between chick's solitary play and under-controlled or aggressive behavior later in life.

Are keel bone fractures in laying hens related to bone strength or to fearfulness?

T. Bas Rodenburg[1], Jasper L.T. Heerkens[2] and Frank A. M. Tuyttens[2]
[1]Wageningen University, Behavioural Ecology Group, P.O. Box 338, 6700 AH Wageningen, the Netherlands, [2]Institute for Agricultural and Fisheries Research, Animal Science Unit, Scheldeweg 68, 9090 Melle-Gontrode, Belgium; bas.rodenburg@wur.nl

The relatively high incidence of keel bone fractures in alternative systems for laying hens is cause for concern. It is still unclear what the major causes for these fractures are. One of the dominant hypotheses is that high fracture rates in non-cage systems are related to high levels of accidents, where birds hit the system with their keel bone during flight and suffer fractures. This effect may be stronger in fearful flocks, where birds may respond hysterically to humans entering the house. On the other hand, there are also indications that bone strength plays a key-role, with birds with weaker bones being more at risk. That would also offer an explanation for the relatively high fracture rates seen in furnished cage systems, where the risk of accidents is reduced. The aim of this study was to compare bone strength and fearfulness in laying hens with and without keel bone fractures from both furnished cages and non-cage systems. To meet this aim 13 commercial flocks, 7 non-cage and 6 furnished cage flocks, were visited around 60 weeks of age. Fifteen birds from each flock, caught at different locations in the house, were tested in a tonic immobility test, with a maximum test duration of five minutes. After all birds were tested, they were killed humanely using electrical stunning or the CASH poultry killer and taken to the lab for dissection. After dissection, it was established whether each bird had a keel bone fracture or not (0/1 trait). Further, from each bird the keel bone, leg bones and wing bones were collected for bone strength measurements (N) using a Versatest™ test stand. Effects of housing system, absence or presence of keel fractures and their interaction on bone strength and fearfulness were tested in a mixed model with flock nested within housing system as random effect. Overall, birds from non-cage systems had a higher incidence of keel bone fractures (87 vs. 65%), a shorter duration of tonic immobility (44 ± 4 vs. 130 ± 9 s; $P<0.001$) and stronger keel (112 ± 4 vs 69 ± 2 N; $P<0.001$) and wing bones (192 ± 5 vs. 155 ± 3 N; $P<0.01$) than birds from furnished cages. Birds with keel fractures had weaker leg bones (176 ± 3 vs. 196 ± 7 N; $P<0.001$) in both systems. For keel strength, birds from non-cage systems with keel fractures tended to have weaker keels than birds without keel fractures (109 ± 4 vs. 134 ± 17 N; $P=0.06$). No difference in tonic immobility was found between birds with or without keel bone fractures. These results indicate that keel bone fractures in laying hens are indeed related to bone strength, but not to fearfulness.

Managing non-beak trimmed hens in furnished cages: using two strains of laying hens and extra environmental enrichment

Krysta Morrissey[1,2], Tina Widowski[1], Laurence Baker[2] and Victoria Sandilands[2]
[1]*University of Guelph, Animal and Poultry Science, 50 Stone Rd E, Guelph, Ontario N1G 2W1, Canada,* [2]*SRUC, Avian Science Research Centre, Auchincruive, Ayr KA6 5HW, United Kingdom; krysta.morrissey@sruc.ac.uk*

Beak trimming (BT) is the most effective and reliable method of reducing feather damage and mortality associated with injurious pecking (IP). Although infrared BT is currently permitted within the UK, this is due for review by the government in 2015, with a view to banning routine BT in 2016. Therefore, this study aimed to identify certain factors that may be useful for successfully housing non-beak trimmed hens in furnished cages. We used a 2×2×2 factorial design to assess the effects of breed (Lohmann Classic (L) or Hyline Brown (H)), beak treatment (trimmed (T) or not (NT)), and environment (extra enrichment (EE) or none (NE)) on mortality, behaviour, and feather condition (FC) throughout the laying period. Extra enrichment included 8 polypropylene ropes, 2 pecking mats (ROWA), and 2 beak blunting boards. At 16 wk of age, birds were assigned to 1 of 8 treatments in a commercial furnished cage facility. Each treatment had 8 replicate 80-bird cages. Behaviour and FC data were collected every 4 wk from 19-71 wk, and mortality was recorded daily. Data were analysed in Genstat using Linear Mixed Models with significance at α=0.05. Residuals were normalized using arcsine +0.375 (percent mortality) and +1 log (FC, rates of behaviour) transformations. Breed had a significant effect on overall mortality (H: 1.5±0.3%, L: 3.0±0.6%; $F_{1,56}$=4.68, P=0.035) as well as cannibalism-related mortality (H: 0.1±0.1%, L: 0.9±0.3%; $F_{1,56}$=4.23, P=0.044). Mean FC per body part was affected by breed × age ($F_{13,725}$=2.11, P=0.012) and beak treatment × age ($F_{13,724.9}$=7.25, P<0.001), as FC for both L and NT hens worsened more quickly over time. L hens tended to perform more total IP ($F_{1,55.8}$=3.83, P=0.055), but no effect of beak treatment was found ($F_{1,55.8}$=2.06, P=0.157). EE did not appear to affect FC ($F_{1,55.4}$=0.63, P=0.43) even though EE hens were observed performing less total IP ($F_{1,55.8}$=5.31, P=0.025). H hens had fewer mortalities and better FC; however, this study was performed in a commercial shed with other non-experimental hens housed in the same facility (all H-T). Therefore, the shed was managed to suit this strain of hen, potentially causing the breed differences observed. Although mortality was not affected by beak treatment, two L-NT cages had to be culled due to high levels of cannibalism (6.25% (EE) and 8.75% (NE)). Hens used the EEs, though their presence did not reduce mortality or feather damage caused by IP. The enrichments may require modification before there are effective as a tool for the successful housing of NT hens.

Changing the animal or the environment: changes in breeding strategy and housing conditions to improve the welfare of pigs

J. Elizabeth Bolhuis[1], Inonge Reimert[1], Winanda W. Ursinus[1,2], Irene Camerlink[1,3], T. Bas Rodenburg[4] and Bas Kemp[1]
[1]*Wageningen University, Department of Animal Sciences, Adaptation Physiology Group, P.O. Box 338, 6700 AH Wageningen, the Netherlands,* [2]*Wageningen UR Livestock Research, Animal Behaviour & Welfare, P.O. Box 65, 8200 AB Lelystad, the Netherlands,* [3]*Wageningen University, Department of Animal Sciences, Animal Breeding and Genomics Centre, P.O. Box 338, 6700 AH Wageningen, the Netherlands,* [4]*Wageningen University, Department of Animal Sciences, Behavioural Ecology Group, P.O. Box 338, 6700 AH Wageningen, the Netherlands; liesbeth.bolhuis@wur.nl*

Interactions between pigs may, apart from the physical environment in which they are kept, profoundly affect their welfare, and are influenced by their genetic background. Behaviour of pigs is, however, difficult to address through breeding. Indirect (or associative, or social) genetic effects (IGE), i.e. heritable effects that individuals have on traits of their group members, might be used to indirectly obtain pigs that behave well in groups. We studied pigs (n=480) diverging in IGE on their pen mates' growth (IGE_g) in 80 barren (B) or straw-enriched (E) pens. We recently found pigs with a favourable IGE_g (IGE_g^+) to inflict less tail damage, show less ear biting and chewing on toys than IGE_g^- pigs. Moreover, after 24-h temporary mixing with unfamiliar pigs, IGE_g^- pigs showed more aggression upon reunion with their own pen mates. These behavioural differences may be associated with or reflect physiological differences. Here, we aimed to study effects of the new breeding strategy and housing on concentrations of leukocytes, lymphocytes and haptoglobin. Blood was drawn at 8, 9 (3 days after mixing) and 22 weeks. Housing, IGE_g class, week and interaction effects were analysed with mixed models. Leukocyte levels decreased over weeks (P<0.001) and E pigs had lower levels than B pigs in week 22 (17.0±0.22 vs. 18.5±0.26·10^9/l, housing × week, P<0.01). IGE_g^+ pigs had lower leukocyte levels than IGE_g^- pigs (17.8±0.2 vs. 18.6±0.3·10^9/l, P<0.05), and a larger decrease after regrouping (-2.5±0.4 vs. -1.6±0.4·10^9/l, P<0.05). Lymphocyte levels decreased over weeks (P<0.01) and were lower in IGE_g^+ than in IGE_g^- pigs (8.6±0.2 vs. 9.2±0.2·10^9/l, P<0.05). Haptoglobin increased after regrouping (P<0.001). E pigs had, overall, lower haptoglobin levels than B pigs (0.57±0.03 vs. 0.66±0.03 mg/ml, P<0.05), and haptoglobin tended to be lower in IGE_g^+ than IGE_g^- pigs (0.58±0.03 vs. 0.65±0.03 mg/ml, P<0.10). No IGE class × housing interactions were found. Similar to the behavioural results, the effects of IGE_g^+ paralleled those of E housing. This suggests that both selection for favourable IGE_g and enrichment reduce (chronic) stress experienced by pigs and could be used in concert to improve their welfare.

Use of evolutionary operation technique on farm level

Heidi M-L. Andersen, Erik Jørgensen and Lene J. Pedersen
Aarhus University, Department of Animal Science, Blichers Allé 20, P.O. Box 50, 8830 Tjele, Denmark; heidimai.andersen@agrsci.dk

To solve health or welfare problems on herd level, farmers try different management initiatives. Often more than one procedure on the farm is changed at the same time without a systematic plan, making it difficult to get an overview of the impact of the different procedures and possible interactions between them. Evolutionary operation (EVOP) is a technique for the systematic experimentation with and improvement of an ongoing full-scale production without actually interrupting it. EVOP use a randomized controlled trial to solve a specific problem, normally by optimizing two or three parameters at a time. EVOP is intended to introduce small changes in the process during normal production. These changes are not large enough to interrupt production, but are significant enough to provide valuable local knowledge about what the optimal herd procedures are. EVOP is used in other areas including biological processes, but has, to our knowledge, not been used in animal production. The purpose of this project was to exemplify the use of EVOP on a pig herd and use the EVOP technique to see how three different management procedures affect the pigs' water intake during the day. As it is assumed, water consumption can be used as a measure of the animals' circadian rhythm. A three factorial experiment with 654 pigs (34.3±4.1 kg) was carried out. The variables were: stocking density (14 or 18 pigs per pen), number of straw allocations (one allocation of 140 g straw per pig/day or four daily allocations of 35 g straw per pig) and allocation of pigs to pens (randomly or by size). The pigs were fed ad libitum. Water use and temperature at pen level were continually measured during the experimental period (28 days). The day was divided into two 12 h periods: 'day' (from 7:00 h to 18:59 h) and 'night' (from 19:00 h to 06:59 h) and the proportion of water use during the night of the total water used was calculated at pen level. Data was analyzed using a linear model. On average, 25.9% of the total amount of water was used during night. Increasing stocking density increased the proportion of water used during night with 2.2 percentage points (P=0.02). Random distribution of the pigs instead of sorting pigs by size at pen level reduced the proportion of water used during night by 2.3 percentage points (P=0.03), while no significant effect of the number of straw distributions (P=0.40) or interactions were found. If the water consumption is a measure of the animals' circadian rhythm, the results indicate that the stocking density and sorting of pigs by size impact on the pig's circadian rhythm. The results show that the use of EVOP may be powerful enough to give a fast indication of the optimal combination of production factors within the herd.

Usage of mechanical brush by Japanese Black calves after early separation from their dams

Ken-ichi Yayou[1], Daisuke Kohari[2] and Mitsuyoshi Ishida[3]
[1]National Institute of Agrobiological Sciences, Tsukuba-shi, 305-8602, Japan, [2]Ibaraki University, Inashiki-gun, 300-0331, Japan, [3]Institute of Livestock and Grassland Science, Tsukuba-shi, 305-0901, Japan; ken318@affrc.go.jp

The early maternal deprivation has long-lasting negative effects on growth and behavioral development in several animal species. Though early cow-calf separation has become popular in dairy and even in beef industry, people are less concerned about the negative effect. We have developed a mechanical brush, an automatic rotating brush, for calf as an alternative for mother's grooming and examined its effects on calves' growth and behavior. Japanese Black calves were raised in individual pens without mechanical brush (Cont; 4 females and 3 males) or in those with the mechanical brush (Bru: 4 females and 5 males) from 1 to 3 days after birth. About 4 months of age, they were introduced into a group of 6 to 10 age-matched calves. Using continuous video surveillance, time lying and time standing during a 24-h period were measured for each calf at 1 and 2 months of age. In addition, the number of self-grooming per 24-h period was determined. The Bru calves continuously utilized the brush until grouping. The average utilization time per 1 day was 25.3±7.3 (SD) min. The daily weight gains (DG) during 2 months after birth were not different between two groups. Although the DG during 3 months after group introduction were not different in female (Cont vs Bru: 0.48±0.33 vs 0.52±0.11), Bru male calves gained more weight than Cont (Cont vs Bru: 0.18±0.11 vs 0.74±0.24) (Student's t-test, p<0.005). The time budget (standing, lying) and the total number of self-grooming was not different between two groups. At 2 months of age of Bru calves, the total number of self-grooming significantly and positively correlated with the utilization time of mechanical brush per 1 day (Spearman's rank order correlation coefficient, r_s=0.92, P<0.001). Although these data suggest a strong motivation of calves for the usage of the mechanical brush, further research is needed to investigate possible benefits of this device.

i-WatchTurkeys: a smartphone application for on-farm turkeys health and welfare assessment

Inma Estevez[1,2], Joanna Marchewka[2], Tatiane T. Negrão Watanabe[3], Roberto Ruiz[2], Alberto Carrascal[4] and Valentina Ferrante[3]
[1]IKERBASQUE, Basque Foundation for Science, Alameda Urquijo, 36-5, 48011 Bilbao, Spain, [2]Neiker tecnalia, Animal Production, P.O. Box 46, 01015 Vitoria-Gasteiz, Spain, [3]Università degli Studi di Milano, Dept. of Veterinary Science and Public Health, Via G. Celoria 10, 20133 Milan, Italy, [4]Daia Intelligent Solutions S.L., GOIEKI, Goierri Agency for Development, 20100, Spain; iestevez@neiker.net

Current advances in mobile technologies open new possibilities for a more precise management of animals in commercial systems. Due to the large number of birds, and the fast turnover of the production cycles in meat poultry, flock health and welfare assessment is particularly difficult. Animal caretakers and veterinarians conduct daily checks to determine flock health status. However, outcomes of the assessment are generally non-quantitative and are based on broad perceptions. Continuous, easy access to reliable, historical and up to date information on the health and welfare status of birds in relation to management and environmental factors, collected in a simple manner, can be an important asset for decision-making in the companies, helping improving birds´ health and welfare and can be a useful tool for external welfare assessors. Within the AWIN FP7 EU project, we have developed a free smartphone app for turkey health and welfare assessment (I-WatchTurkey) that in simple steps allows a standardized data collection of critical health and welfare indicators, based on the transect methodology. This app includes the possibility of recording mortality and the frequency of birds (per transect) with relevant welfare and health deficiencies, such as inmobile individuals, severe lameness, wounds, unwanted behaviors or serious health issues in the flocks, without catching or disturbing the birds. Incidence of the health and welfare issues detected are automatically standardized (to percentages) by the number of birds in the flock at the time of assessment and the number of transects conducted. Data collection for assessment can be conducted at the time of a regular health inspection. In addition, the system records automatically the farm geographical location, weather conditions at the time of inspection, and allows entering relevant information, such as birds strain, birds age, housing and management conditions, expanding significantly the applications of the app and data analysis. This tool will allow farmers and veterinarians to screen the performance of their flocks over time, motivating them for self-evaluation and providing better welfare to their animals. The main asset of this innovative solution is its simplicity of data entry, the option to get the report immediately after assessment and it is freely available.

Electronic feeding stations for ewes

Stine Grønmo Vik, Tor Gunnarson Homme and Knut Egil Bøe
Norwegian University of Life Sciences, Department of Animal and Aquacultural Sciences,
Arboretveien 6, 1432 Ås, Norway; stine.vik@nmbu.no

Concentrate given to ewes that are housed during the winter season is usually provided in a feeding trough where all animals eat simultaneously. However, similar to dairy and pig production, electronic feeding stations (EFS) are becoming more common on commercial sheep farms. The maximum number of ewes per feeding station is a major factor in keeping costs low. The aim of this study was to investigate the use and capacity of EFS on commercial sheep farms. The study was conducted on four commercial farms in Norway that had EFS for ewes. The number of animals per EFS were 36, 70, 72 and 80, and the rations of concentrate provided per day were 6, 10, 10 and 3, respectively. The number of displacements and the number of ewes which queued behind the EFS was scored by an observer for three hours in the morning and three hours in the afternoon on each farm. The time of each visit to the EFS, the identity of the ewe and whether the ewe obtained any concentrate feed (rewarded or unrewarded) for three consecutive days were retrieved from the EFS computer. All ewes used the EFS regularly. Total number of daily visits to the EFS varied from 739 to 1.428 between herds and total occupation time per 24 h period varied from 9.3 to 16.8 hours. The number of rewarded visits per ewe per day varied from 3.2 to 5.9, whereas the number of unrewarded visits ranged from 6.0 to 21.5 per ewe per day between herds. Interestingly, the farm with the lowest number of visits to the EFS was the farm with the highest number of ewes in the group (80 ewes). This farm also provided only three concentrate rations per day. The number of displacements from the EFS was low for all the farms. The average number of ewes queuing behind the EFS ranged from 2.8 to 4.5. As all ewes used the EFS regularly, the EFS system is believed to work well for sheep. The number of unrewarded visits and total occupation times were generally high and the data indicate that reducing the number of rations per day may increase the capacity of the EFS.

Use of GIS to analyses cattle transport has potential to improve carbon footprint and animal welfare

Tâmara Duarte Borges[1], Adriano Gomes Páscoa[2], Janaina Da Silva Braga[1], Arquimedes José Riobueno Pellecchia[1], Antoni Dalmau[3] and Mateus José Rodrigues Paranhos Da Costa[4]

[1]Universidade Estadual de São Paulo – UNESP, Programa de Pós-graduação em Zootecnia, Prof. Paulo Donato Castellane, s/n, 14.884-900, Jaboticabal, SP, Brazil, [2]BEA Consultoria e Treinamento, Rua Francisco Puzzoni, 476, 14.882-125, Jaboticabal, SP, Brazil, [3]Research and Technology Food and Agriculture – IRTA, Veinat de Sies s/n. Monells, 17121 Girona, Spain, [4]Universidade Estadual de São Paulo – UNESP, Departamento de Zootecnia, Pesquisador Cnpq, Prof. Paulo Donato Castellane, s/n, 14.884-900, Jaboticabal, SP, Brazil; tamaratdb@hotmail.com

The aim of the study was to test whether the use of geographic information systems (GIS) can be used to improve carbon footprint and animal welfare during cattle transportation from farms to slaughterhouses. Cattle transports were assessed in 3 slaughterhouses plants (P1, P2 and P3) located in South Eastern region of Brazil. Survey was carried out for 1 year and 4 months, recording data from distance (n=61,132) and time spend in cattle transport (n=18,993). The counties of origin of these animals and the geographic coordinates of slaughterhouse plants were mapped using a geographic identifier reference provided by Brazilian Institute of Geography and Statistics using a correlation in Idrisi v15 Andes program. The means distances travelled per journey were, on average, similar for the 3 slaughterhouses plants (207.8±197.4, 223.0±122.3 and 190.4±171.7 km/journey, for P1, P2 and P3, respectively). Shortest distances were about the same for the 3 plants, 40 km/journey, lasting, on average, only 30 min/journey. The maximum of journeys length and duration were more variable among the plants, with 1008 (23.5 h/trip), 658, (8 h/trip), and 976 km (22 h/trip) for plants P1, P2 and P3, respectively. Distances longer than 600 km, represented only 8.4, 0.92 and 1.9% of the journeys assessed in P1, P2 and P3, respectively; but proportions were different when we consider journeys with more than 6 hours (9.5, 14.4, and 5.1% for P1, P2 and P3). It is assumed that long distances and long transportation periods potentially increases the risks of impoverishing animal welfare and carbon footprint of beef cattle chain. Part of this assumption was confirmed empirically, with data from 210 journeys with a GPS fitted in the trucks and the percentage of carcasses with bruising; results showed that long transportation periods (longer than 6 hours) had a negative effect on cattle welfare, increasing 0.2 bruising/carcass per extra hour after the first six (P<0.05). We concluded that with the application of GIS tools it is possible to offer valuable information about where to purchase cattle, and develop a logistics strategy that minimize risks of damaging animal welfare, beef quality and environment.

Positve effects of the good practices of handling adoption on the welfare of dairy calves

Lívia Carolina Magalhães Silva, Luciana Pontes Da Silva, Maria Fernanda Martin Guimarães and Mateus José R. Paranhos Da Costa
Faculdade de Ciências Agrárias e Veterinárias, UNESP, Zootecnia, Via de acesso Prof. Paulo Donato Castellane s/n, 14884900, Brazil; lmagalhaesilva@gmail.com

The aims of this study were to assess the effects of keeping dairy calves with their dams for the first 24 hours of life (CD) and of to stimulate them tactily during suckling (TS) on their welfare. Forty eight dairy calves from Girolando breed were distributed in 4 treatments (T): T1 = CD without TS; T2 = CD with TS; T3 = calves were separated from their dams soon after birth, and the first care carried out by the handler (CH) without TS and T4 = CH with TS. The calves of all treatments received colostrum within 3 hours after birth and were handled in the same way from birth to 120 days of age. Animal welfare was assessed using behavioural indicators, with an unknown person applying two test, the Flight Distance Test (FD, measuring the minimum tolerated approach distance by a human in a yard) and Docility Test, by measuring three variables: the latency to the calf to present the first movement (LM), the length of time to hold each calf in a corral corner (LTH), and the total time touching the calves (TTT), in seconds. The analyses of variance were carried out with repeated measures, adopting the method of restricted maximum likelihood (REML) and the MIXED procedure from SAS statistical software package, considering the fixed effects of treatments, ages (30, 60, 90 and 120 days of age), and the interactions between them. The interactions affected significantly FD (F=2.08, P<0.05), LTH (F=12.57, P<0.001) and TTT (F=5.00, P<0.001), and LM was affected only by the treatments (F=7.35, P<0.001), and the T3 had the highest mean (7.50 ± 0.65 s) and T2 the lowest (4.58 ± 0.67 s), indicating that LM could be related to the fear towards humans, when the first movement is towards humans. In the first two assessments (30 and 60 days of age) calves from T2 and T4 had lower DF means than T1 and T3. From 90 days of age there was a significant increase of DF for T4 calves, suggesting that the absence of the mother in the first 24 hours of life can reduce the fear response of calves towards humans. T4 calves showed lower means of LTH at 30 and 60 days of life, and the tactile stimulated calves (T2 and T4) presented longer TTT, followed by T1 and T3 calves, respectively. T3 calves had a lower TTT means in all ages. The results of behavioural tests confirmed the expectations that tactile stimulated calves (T2 and T4) would be less fearful towards humans and easier to be handled than those without stimulation. Besides, the mothers absence in the first 24 hours of life seemed to facilitate calves bonding with humans. Financial support: FAPESP (Process number 2011/00388-3).

Behavioral response to mastitis challenge from Holstein dairy cows in early lactation

Peter D. Krawczel[1], Gina M. Pighetti[1], Raul A. Almedia[1], Susan I. Headrick[1], Lydia J. Siebert[1], Mark Lewis[2], Charlie Young[2], Randi A. Black[1] and Stephen P. Oliver[1]
[1]*The University of Tennessee, Department of Animal Science, 2506 River Drive, Knoxville, TN 37996-4574, USA,* [2]*The University of Tennessee, East Tennessee Research and Education Center, 3209 Alcoa Hwy, Knoxville, TN 37996-4576, USA; krawczel@utk.edu*

The objective of this trial was to determine the behavioral changes that occur in early lactation dairy cows, with or without prior exposure to an experimental Streptococcus uberis vaccine, following an intramammary challenge with this organism. Twenty-five Holstein dairy cows, either treated with an experimental vaccine (n=11) or saline control (n=14) were enrolled on this study. Intramammary challenges with S. uberis were conducted 1.2±0.7 days after calving and data were collected from day -2 to day 6, relative to challenge. Lying time (h/d), lying bouts (n/d), lying bout duration (min/bout), and steps (n/d) were recorded at 1-min intervals using dataloggers. Data were analyzed with a mixed procedure in SAS (v9.2) to establish effects of vaccination status, day and their interaction. Cow within treatment was used a repeated measure. Following challenge, all cows developed clinic mastitis, as defined by presence of S. uberis in a minimum of 2 milk samples collected from challenged quarter, increased somatic cell counts, increased milk scores, and increased udder scores. Independent of vaccination status, lying time increased (P=0.02) from 8.7±0.5 h/d on d 0 (administration of challenge) to 10.2±0.5 h/d on d 6. Vaccinated cows tended to spend fewer hours lying per day (1.2±0.4 h; P=0.06) compared to the control. Lying bouts differed (P=0.02) relative to day of challenge. Lying bout duration did not differ between treatments (P=0.67), but increased (P<0.001) from 69.8±6.0 min/bout on d 0 to 81.4±6.0 min/bout on d 6. Steps per day did not differ between treatments (P=0.93), but decreased (P=0.02) from 2,357±178 n/d on d 0 to 1577±178 n/d on d 6. These data suggest that our experimental vaccine did not provide protection from mastitis infection as all cows meet our criteria for clinic mastitis following challenge. The consistent decrease in overall activity following challenge suggests the increasing commonality of automated devices for monitoring behavior on commercial dairy farms may provide a means to improve mastitis detection.

Outdoor areas for dairy cows – behaviour, welfare and production

Lise Aanensen, Svein Morten Eilertsen and Grete H.M Jørgensen
Bioforsk Nord Tjøtta, Parkveien 1, 8860 Tjøtta, Norway; lise.aanensen@bioforsk.no

From 01.01.14 all Norwegian dairy farmers are obliged to give their dairy cows, including cows in loose housing, access to pasture for minimum 8 weeks per year. Dairy barns are often located far from pastures and the herds are large. Farmers and the dairy industry seek solutions for the technical and practical challenges. Can a simple exercise pen replace access to pasture? In 2013, we performed a pilot study to investigate how dairy cows use different outdoor areas in two commercial dairy farms with loose housing and automatic milking (AMS). Farm 1: fifty dairy cows, DeLaval milking robot, guided cow traffic. 2.8 ha green pasture for 33 days. Farm 2: fifty dairy cows, Lely milking robot, free cow traffic. 0.7 ha exercise enclosure in a small forest area for 15 days. Outdoor access during daytime. Registered weather type and number of cows outside. Within farms we compared activity, behaviour and milk production on days with and without access to outdoor areas. Farm 1. In average 66% of the cows went outside when possible. Cows exhibited significantly higher activity on 'pasture-days'; standing/walking 80% of the time, lying 20% of the time, walking in average 275 steps per day (9 am - 4 pm). 72.2% of all observations were spent grazing, only 4.7% of the observations were lying. On 'indoor-days', cows were standing/walking 37% of the time, lying 63% of the time and walking 70 steps (9 am-4 pm). Milk yield was however lower (P<0.01) and number of visits to the AMS fewer (P<0.001), on 'pasture-days'. Fewer cows went out on rainy/drizzly or bright sunny / hot days compared to cloudy days (P<0.05). Farm 2. In average 31% of the cows went outside when possible. Access to the outdoor area gave no significant effect on cow activity (P=0.8). Main outdoor observations; standing/walking with head up (43.2%) and lying (33.2%). Even though they only had access to a small forest area, eating counted for 18.3% of the observations. The number of cows outside was mainly controlled by the indoor feeding interval and not by the weather (P=0.115). Access to the exercise enclosure did not affect the daily milk yield, but there was an increase in number of visits to the AMS (P=0.005). In conclusion, dairy cows went outside if possible. The number of cows and time spent outside depended however on feed access and weather conditions. Access to green pasture increased cow activity, and reduced milk yield and AMS visits, while access to an outdoor exercise enclosure had no effect on cow activity but gave more visits to the AMS. Access to outdoor areas increased the available space, which in turn gave subordinate individuals the oportunity to avoid social conflicts and increased access to resources. Consequently, this will have a positive effect on animal welfare.

Temperature responses in Holstein lactating cows fed with different forage: concentrate ratio under heat stress condition

Jalil Ghassemi Nejad[1], ByongWan Kim[1], Jayant Lohakare[2], Joe West[3], BaeHun Lee[1], DoHyun Ji[1], JingLun Peng[1] and KyungIl Sung[1]

[1]*Kangwon National University, Animal Life System, KNU Ave. 1, 200-701 Chuncheon, Gangwon Province, 200-701, Korea, South,* [2]*Kangwon National University, Animal Biotechnology, KNU Ave. 1, 200-701 Chuncheon, Gangwon Province, 200-701, Korea, South,* [3]*Georgia University, Animal and Dairy Science, Tifton campus, 31793-0748 Tifton, GA 31793-0748, USA; jalil@kangwon.ac.kr*

Measurement of body temperature is a common method for an evaluation of cow's well-being which is currently used on dairy farms. Body temperature may vary due to different factors one of which is high ambient temperature. Fever can impair animal welfare usually occurs in thresholds of 39.4 °C to 39.7 °C. Higher forage ratio may exacerbate the body temperature increase due to producing higher heat increment and more rumen activity. Therefore, temperature responses in heat-stressed Holstein lactating cows fed with different forage:concentrate ratios were evaluated using 20 Holstein lactating cows (10 cows per treatment).This study was conducted during July and September, 2012 for 64 d. The experiment was carried out at Naju, republic of Korea. Completely randomized design was performed and data were analyzed using proc ANOVA of SAS. The forage:concentrate (F:C) ratio in control and treatment groups were 48:52 and 56:44, respectively. Lactating cows were housed in sheltered drylot facility with ad-libitum access to water. The highest daytime temperatures reached around 35 °C and the relative humidity was about 86%. Body and skin (surface) temperature of each cow in 7 points of the body including rectal, vagina, hip, udder, rumen-side (flank), ear, and forehead were measured using infra-red gun (having two mode, body and skin). At the starting point of the experiment no significant differences were observed in bodys' and skins' temperature of these 7 points of the cow. But, at the end of experiment vagina's body temperature was higher (P=0.041) and rectal temperature tended to be higher (P=0.083) in treatment group compare to control group. Ear's body temperature was lower and ear's skin temperature was higher (P=0.032) in treatment group compare to control group. Forehead body temperature was higher (P=0.048) in control group than treatment group while forehead's skin temperature was lower in treatment group (P=0.041). No differences were observed in hip, udder and rumen-side (flank) temperature (both in body and skin) between treatment and control group (P=0.012). It is concluded that temperature of different parts of body of lactating cows may change during heat stress while fed with different forage:concentrate ratio. More researches are needed to investigate the regulation of body temperature with different F:C ratio in cattle under heat stress.

Effect of mineral mixture on behavior patterns of Portuguese native heifers during Mediterranean spring and summer

Lara Prates[1], Teresa Dias[1], Muirilo Quintiliano[1], José Castro[1], Ana Geraldo[1], Cristiane Titto[2], Evaldo Titto[2], Paulo Infante[1] and Alfredo Pereira[1]

[1]University of Évora, ICAAM – Institute of Mediterranean Agricultural and Environmental Sciences, Herdade da Mitra, Apartado 94, 7000-554, Portugal, [2]Faculdade de Zootecnia e Engenharia de Alimentos, Universidade de São Paulo, Medicina Veterinária, Av. Duque de Caxias Norte, 225, Campus da USP, CEP 13635-900 Pirassununga/SP, FZEA, 13635-900, Brazil; apereira@uevora.pt

The aim of this study was to compare, on spring and summer, the result of the availability of mineral mixture with non-protein nitrogen, on the behavior and performance of Portuguese native's heifers. The experiments lasted for 147 days and were used 10 heifers. The animals remained in two contiguous parcels of natural grassland with approximately 3.5 ha and 4.0 ha of the area, without mineral supplementation (NS) and with supplementation (WS), respectively. The study was design in 2 periods, spring and summer: in the spring (P1) the animals had natural pasture and in the summer (P2) had dry pasture supplemented with hay. The behaviour's observations were focal, each five minutes, from 7 am to 8 pm. Regarding the behavior daily patterns, there were no significant changes resulting from mineral supplementation. However, the influence of high temperatures determined changes in the behavioral patterns and on the distribution of activities throughout the day. The time spent on food intake was the longest activity in both seasons (P1-NS 517 min.; P1-WS 527 min; P2-NS 248 min and P2-WS 162 min), rumination (P1-NS 65 min.; P1-WS 79 min; P2-NS 93 min and P2-WS 85 min) and resting (P1-NS 193 min.; P1-WS 161 min; P2-NS 158 min and P2-WS 158 min). The water consumption daily frequency was significantly different between periods (P1-NS 1.6 times; P1-WS 1.73 times; P2-NS 2.16 times and P2-WS 3.1 times). In group WS, the daily search for mineral salt with non-protein nitrogen was higher in the summer due to the poor quality of available food (2 times/day in P1 vs 4,73 times/day in P2). Mineral supplementation caused positive effects on animal's performance, lessening the decrease in average daily gain during the summer (P1-NS 730 g/d; P1-WS 760 g/d; P2-NS 430 g/d and P2-WS 510 g/d) and in the higher concentrations of thyroxine. In conclusion, it can be stated that heat stress influenced the grazing patterns. The combined effect of quality of the food and the differences in environmental temperatures seem to be responsible for the lower average daily gains during the summer relative to spring. The protein and mineral supplementation significantly improved the performance. These results emphasized the role of the mineral mixture with non-protein nitrogen supplementation on Mediterranean environment, particularly during the summer period.

Behavior of replacement heifers in hot temperatures depending on availability of shade

Corrie V. Kerr[1], James Reynolds[1], Stephen C. Adolph[2] and Jose M. Peralta[1]
[1]Western University of Health Sciences, College of Veterinary Medicine, 309 East Second Street, Pomona, CA 91766, USA, [2]Harvey Mudd College, 301 Platt Boulevard, Claremont, CA 91711, USA; ckerr@westernu.edu

Increased heat load caused by hot weather can have strong adverse effects on animal physiology and well-being. Certain environmental factors, including solar radiation, can contribute to increased heat load. Shade can reduce an animal's radiant heat load by 45% and is typically provided on dairy farms for calves and cows due to concerns regarding their vulnerability. However, less is known about how shade affects heat stress on growing heifers, and thus shade is usually not provided for them. The aim of this study was to investigate whether heifers prefer shade and the environmental conditions that determine when shade becomes important for thermoregulation. The study, conducted at a California dairy during July 2013, monitored shade preference in 86 Holstein heifers divided into three age groups (3, 14 and 22 months). Two treatments (no-shade and shade) were provided for each group. Specific behaviors were recorded for 14-15 cattle in each group and treatment type, and behavioral analysis compared thermoregulatory responses between animals with and without shade access. Preference for shade increased across all age groups as heat load increased. An increase in shade use paralleling temperature-humidity index was noted among the two older groups. Black globe and effective temperature values were significantly higher in the sun than in the shade. Thermoregulatory behaviors were observed with greater frequency and in more animals in the no-shade treatments, especially in those with darker coats. In conclusion, heifers seek shade in hot environments and those without shade exhibit behaviors indicative of heat stress. This study supports the thermoregulatory response of dairy heifers to seek protection from the sun on hot days. Re-evaluation of existing guidelines for the housing of growing replacement heifers may help to better accommodate animals in climatic regions with high temperatures. Ultimately, the provision of shade can enhance the welfare of heifers worldwide.

Changes in the lying behaviour of pre and post-partum dairy goats

Gosia Zobel[1], Ken Leslie[2], Daniel M. Weary[1] and Marina A. G. Keyserlingk[1]
[1]*University of British Columbia, Animal Welfare Program, 2357 Main Mall, Vancouver, BC, V6T1Z4, Canada,* [2]*University of Guelph, Department of Population Medicine, Ontario Veterinary College, 50 Stone Road, Guelph, ON, N1G2W1, Canada; g_zobel@yahoo.ca*

Lying behaviour often changes in response to management, housing and illness, and these changes are commonly used as a measure of welfare in dairy cattle. To our knowledge, no work has examined how lying behaviour changes before and after parturition in dairy goats. The objective of this study was to describe the lying behaviour of commercially housed dairy does in the last two weeks of pregnancy and the first week following kidding. Electronic data loggers were attached to the rear left leg of 36 Alpine crossbred does (entering third lactation). Does were housed in the same pen before and during kidding. Kids were removed within 4 hours of kidding. Does were moved to the same milking pen after their first milking. Daily lying time and duration of lying bouts was calculated per doe. Since lying behaviour was similar across does (14.5±0.4 h/d) during the two weeks before kidding (d-14 to d-1), this pre-partum period was used for comparison with post-partum data. Three periods were used for comparison: day of kidding (d0), the day after kidding (d1) and the week following kidding (d2-7). Does carrying single kids lay less pre-partum than does carrying multiple kids (13.0±0.6 vs. 14.8±0.3; P=0.03), therefore number of kids was included in the analysis as a covariate. Mixed model analysis was conducted for lying time and lying bout duration with period as a repeated measure. Results are presented as means ± SE. On day of kidding (d0), does decreased their lying time by over 50% (mean lying time = 6.9±0.6 h/d; P<0.0001). On the day following kidding (d1), does increased lying time but continued to lie down less compared to pre-partum (11.1±0.5 h/d; P<0.0001). In the days following (d2-7), lying time increased but remained lower than pre-partum levels (13.6±0.3 h/d; P=0.02), in part due to twice daily milking activities. Similar to lying time, the duration of individual lying bouts also decreased on d0 when compared to the pre-partum period (20 vs. 36±2 min/bout; P<0.0001), indicating that does were more active on the day they gave birth. Overall, these results provide the first evidence that lying behaviour of dairy does changes dramatically in the days around kidding. The large change in lying bout duration and daily lying time may be attributed to physical changes associated with the birthing process, or as a response to the removal of kids (a common management practice). Monitoring lying behaviour could be used to detect impeding parturition, and to identify abnormal behavioural patterns indicative of illness or other post-partum issues.

Development of a protocol to assess the welfare of sheep

Roberto Ruiz[1], Ignacia Beltrán De Heredia[1], Josune Arranz[1], Susan Richmond[2], Francoise Wemelsfelder[2], Elisabetta Canali[3] and Cathy Dwyer[2]
[1]NEIKER-TECNALIA, Arkaute Centre, 01080 Vitoria-Gasteiz, Spain, [2]SRUC, Roslin Institute, EH 259RG, Easter Bush. Scotland, United Kingdom, [3]University of Milan, Via Celorio, 10, 20133 Milano, Italy; rruiz@neiker.net

There is a wide diversity of sheep production systems in Europe, most of them fitted to the seasonal utilisation of available natural resources. Sheep can be kept indoors during certain periods of the year, or graze outdoors. This variability in management and rearing conditions determines the productivity but also the behaviour and welfare of the animals. The development of an animal-based protocol to assess the welfare status of the sheep should encompass the diversity and seasonality in rearing conditions. Within the AWIN project, a two-step protocol has been designed to rapidly screen farms for welfare in stage 1, to follow up with a detailed second stage if welfare concerns are detected. This prototype is being tested twice a year (winter and summer) in commercial flocks representing the diversity in farming conditions (more or less grassland based), geographical locations, and production type (dairy and meat). At stage 1, a sample of between 40 to 100 ewes (depending on the number of individuals within a flock) is evaluated from a distance without disturbing the animals. Group animal-based assessments include: (1) QBA (qualitative behavioural assessment), (2) quantitative behavioural assessment: such as the proportion of ewes standing, lying, panting, scratching and vigilant, (3) physical status: coat cleanliness, fleece length and quality, tail length and incidence of lameness, (4) fear assessment: flight distance to the approach of humans and startle responses. In addition, a resource-based assessment of the environmental conditions (water availability, shelter, food or pasture availability, landscape, topography, etc.) is recorded to supplement animal-based indicators. At stage 2, individual assessment of 50 to 150 ewes is performed, depending on the flock size also. Breed, age and reproductive status, day and time of assessment and weather conditions are recorded together with an individual physical assessment of all ewes in the sample including: teeth status, mucosa colour, eyes (evidence of injury, infection or disease), nasal discharge, ears (injuries, damage from tags, numbers of tags, etc.), horns (polled, trimmed, or damaged); respiration, skin lesions, fleece length and quality, coat cleanliness, evidence of lesions or calluses on the legs; body condition, udder and teats conditions, dag score, tail length and lameness. The current status of this prototype and the preliminary results of the implementation of the test during the winter months on the monitored flocks will be presented.

Assessing inter and intra-observer reliability in animal-based welfare indicators in dairy sheep

Ignacia Beltrán De Heredia[1], Josune Arranz[1], Inma Estevez[1], Susan Richmond[2], Francoise Wemelsfelder[2], Elisabetta Canali[3], Cathy Dwyer[2] and Roberto Ruiz[1]

[1]NEIKER-TECNALIA, Arkaute Centre, 01080 Vitoria-Gasteiz, Spain, [2]SRUC, Roslin Institute, EH 259RG, Easter Bush. Scotland, United Kingdom, [3]University of Milan, Animal Husbandry, Via Celorio, 10, 20133 Milano, Italy; ibeltran@neiker.net

Sheep welfare assessment presents particular challenges related to the exposure to environmental conditions and to the seasonality of the breeding cycle. To assess the feasibility and limitations of an animal-based protocol for welfare indicators in dairy sheep, 30 ewes were randomly chosen according to the age structure of the flock of Neiker. The selected ewes represented 23% of the flock, their age ranging from 1.5 to 8.5 years old. The animals were assessed six times during a productive cycle (July 2012 to June 2013). For each assessment, the ewes were gathered, identified and scored according to the welfare indicator protocol developed within the EU-FP7 'AWIN' project. Indicators assessed the general health status, sanitary conditions and lesions and injuries. A blood and a faecal sample were collected from each ewe to assess general blood parameters and parasite burdens. To test inter and intra-observer variability, the rate of agreement within an observer in two consecutive rounds and between three observers were assessed. The resulting database was analysed by a mixed model (Proc GLIMMIX, SAS) with Observer (O) as a fixed effect, the day (D) and the round (R) as double repeated measures. Spearman correlations were conducted to determine the relationship across parameters. Both O and D had a significant effect on body cleanliness, dag scoring, BCS and mucosa color ($P<0.001$). R and D were also significant for body cleanliness ($P<0.001$), as well as O*R for BCS and body cleanliness. Regarding the effect of D, sheep showed poorer conditions during final pregnancy and middle lactation, and it appeared to be related to the environmental conditions and managing procedures. There was lack of variability for many of the indicators related to general health status and sanitary conditions with a low incidence. Blood biochemical parameters were within the normal ranges throughout the whole cycle. As for parasites, Dicrocoelium and Strongilides were found in every control, but counts were below 100 eggs/g faeces. No significant correlations were found across apparently related parameters, such as dag scoring and parasites, dag scoring and BCS, color of mucosas and haemoglobin, milk yield and SCC, etc. ($P>0.05$ in all cases). These results show that, within a single farm with consistent management, significant seasonal variation occurs. To overcome the differences between observers, more objective thresholds are required and the unification of criteria for some of these indicators.

Transport density and lamb's welfare

Evaldo Titto[1], Thays Leme[1], Cristiane Titto[1], Alfredo Pereira[2], Saulo Silva[1], Ana Geraldo[1], Reissa Vilela[1], Roberta Sommavilla[1] and Raquel Calviello[1]
[1]Animal Science and Food Engineering College, Biometeorology and Ethology Laboratory, Av. Duque de Caxias Norte, 225, Pirassununga – SP, 13635-900, Brazil, [2]University of Evora, Biometeorology and Animal Welfare Laboratory, Departamento de Zootecnia, Universidade de Évora, Pólo da Mitra, Apartado 94 7002-554 Évora, Portugal; titto@usp.br

Transport is a very important economic factor because it can cause physical and physiological stress in farm animals, beyond negative effects on health, welfare, performance and the final product quality. High or low density during transport can cause problems or injures and falls during transport. For this reason, the number of animals and the space available in the truck must be considered. The aim of this study was to investigate the effects of loading density on blood cortisol variation. Sixty-four White Dorper lambs were transported once for 4 h (195 km) to a commercial slaughterhouse using completely closed and bedded truck. Lambs were located in two separate truck compartments according to loading density: 0.3 m^2/animal (n=28; 14 female, 14 male) and 0.2 m^2/animal (n=36; 13 female, 13 male). The air circulation was from the top of the cage, without external visual access. Blood samples were also collected before and after transport for analysis of serum cortisol and samples were taken through jugular venipuncture in Vacutainer® tubes and kept under refrigeration. Data analysis used mixed procedure of SAS using transport density as fixed effect and as random effect the sex, and the interactions density vs. sex. Each lamb was considered as the experimental unit. In case of significant results (P<0.05); it was adopted the Student t test for multiple comparisons. Lambs transported in density of 0.3 m^2 per animal had lower levels of cortisol, when compared to those transported at higher density (19.7 vs. 22.3 ng/ml; P<0.05). The lambs in the largest area had lower variation in cortisol levels compared to the lowest area between before and after transport (-3.7 vs. 9.3%; P<0.05), which can demonstrate greater comfort and a positive influence in reducing animal stress.

Response of unweaned and weaned lambs to social and maternal separation

Eric Romero[1], Francisco González[1], Alfredo Medrano[1], Alberto Tejeda[2], Alan Olazabal[1] and Rosalba Soto[1]
[1]Facultad de Estudios Superiores Cuautitlán, UNAM, Ciencias Pecuarias, Carretera Cuautitlán Teoloyucan, Km 2.5, Cuautitlán Izcalli Edo. de México CP 54714, Mexico, [2]Facultad de Medicina Veterinaria y Zootecnia, UNAM, Etología, Ciudad Universitaria, Distrito Federal, CP 04510, Mexico; rosaneopri@yahoo.com.mx

The response to separation of the mother or of conspecifics in the lamb is not known in detail. The artificial weaning involves the breakup of the mother-young bond at ages that it could be a strong agitation and distress in lambs. Therefore a standardized test was applied in 40 unweaned lambs (45, 60 days old) and 20 weaned (90 days old). Lambs were observed for 15 min in a 2×2 m pen (5 min with their mother and conspecifics; 5 min with their mother in absence of conspecifics and 5 min without their mother but with conspecifics). The conspecifics were the rest of the lambing flock. An agitation index was built using the standardized data of (1) frequency of high-pitched bleats, (2) locomotor activity, (3) attempts at jumping out of the testing pen and (4) eliminations behaviors. The results revealed that in the three assessed ages, the agitation index was lower in social than in maternal separation (0.4±0.3 vs 4.5±1.1; 1.4±0.6 vs 4.2±0.5; 2.0±0.7 vs 5.1±0.6, P<0.01, for 45, 60 and 90 days respectively). However, there was no difference between unweaned or weaned lambs regarding the agitation level observed at the social and maternal separation (P>0.05). Overall, results suggest that lambs undergo more agitation when their mother is removed than when their conspecifics are removed of their social context. It is concluded that until 90 days, the social bond of the lamb toward its mother does not show any disturbance, neither exists a social preference for the conspecifics. The rupture of the mother-young bond by an artificial weaning could impair the welfare of the lamb.

Can providing dustbathing substrates in enriched cages for laying hens help to control mite infestations?

Giuseppe Vezzoli and Joy Mench

University of California, Davis USA, Department of Animal Science and Center for Animal Welfare, One Shields Avenue, Davis, CA 95616-8521, USA; gvezzoli@ucdavis.edu

A presumed function of dustbathing behavior is to remove ectoparasites. The provision of dustbathing substrates in enriched cages for laying hens might therefore offer an alternative to using pesticides to control ectoparasites. We investigated the effectiveness of dustbathing substrates for controlling Northern fowl mites in individually caged beak-trimmed White Leghorn hens (n=32). Each cage contained a 32×32 cm plastic tray that was either: (1) filled with 1,200 g of sand (SAND); (2) empty (CONTROL); (3) covered with AstroTurf (AT); (4) or covered with AT on to which 150 g of feed was delivered daily (ATF). AT and ATF were evaluated because of their use in the dustbathing/foraging area of many newer commercial enriched cages. Hens were infested with approximately 35 mites at 25 weeks of age. Mite populations were visually estimated weekly using a 0-7 scale, with 7 being ≥10,000 mites. Time spent dustbathing both in the tray and on the wire cage floor were video recorded immediately before and after infestation and at 1, 3, 5, and 7 week post-infestation for 2 consecutive days from 12:00 to 20:00 h. Data were analyzed using a repeated-measures ANCOVA in SAS; dustbathing data reported are back-transformed means. Substrate did not influence the total time spent dustbathing (average across substrates: 11.3 min). However, there were substrate effects on the time spent dustbathing in the trays ($F_{2,21}$=3.61, P=0.045) and on wire ($F_{2,21}$=6.96, P=0.005). SAND spent more time dustbathing (11.4 min) than AT (2.4) in the trays, and CONTROL spent more time (11.5) than ATF (0.5) on wire. SAND hens never dustbathed on wire, and CONTROL hens never dustbathed in the empty trays. There was a substrate effect on mite numbers ($F_{3,28}$=3.72, P=0.02), with ATF having more mites (mean score = 5.3±0.27) than AT (4.2±0.27), and with SAND (4.5±0.27) and CONTROL (4.4±0.27) having intermediate scores. There was no relationship between mite numbers and the time spent dustbathing. This study confirmed that the substrate type provided affected dustbathing behavior. However, even though SAND was a preferred dustbathing substrate it was not effective for controlling mite numbers, nor was the time spent dustbathing influenced by infestation levels. In addition, our data suggest that the use of ATF in enriched cages might lead to increased mite numbers in infested hens, possibly because the fat content of the feed contributes to an increase in feather lipids, thus creating a better habitat for the mites.

Access to cooling perches affects the behavioral responses of laying hens during acute heat stress

Maja M. Makagon[1], Patricia Y. Hester[1], Giuseppe Vezzoli[1], Richard S. Gates[2], Stacey A. Enneking[3] and Heng-Wei Cheng[3]

[1]*Purdue University, Animal Sciences, 125 S. Russell St., West Lafayette, IN, 47907, USA,* [2]*University of Illinois, Agricultural and Biological Engineering, 1304 W. Pennsylvania Ave., Urbana, IL, 61801, USA,* [3]*USDA-ARS, Livestock Behavior Research Unit, 125 S. Russell St., West Lafayette, IN, 47907, USA; mmakagon@purdue.edu*

In addition to compromising laying hen welfare exposure to heat stress has deleterious impacts on economically relevant production traits, including hen mortality, egg production, egg weight, shell thickness and body weight. The effectiveness of a cooled perch system for helping hens cope during episodes of heat stress was evaluated. Sixteen-week-old hens were randomly assigned, in groups of nine, to cages furnished with feeders, drinkers, and a standard round galvanized steel perch (SP, n=6), a round galvanized steel perch filled with chilled water (CP, n=6), or no perch (NP, n=6). The cages were arranged in three tiers of two. Cage nested in tier served as the experimental unit. At 27 weeks of age, the hens were subjected to an acute heat stress event during which the ambient room temperature was increased to above 32 °C (range: 32-34.6 °C) for a period of four hours. Hen behaviors were evaluated using live observations before (10:15 h), during (11:15 and 12:15 h), and after (15:15, 16:15, 18:00, 23:00 h) the event. Two instantaneous scans of each cage were made at 15 minute intervals at each time point. Behavioral observations conducted three days prior to the heat stress event (at 10:00 h, 1400 h, 18:00 h and 23:00 h) were used as a baseline (maximum room temperature = 28 °C at 18:00 h). Data were analyzed using the GLIMMIX procedure in SAS and focused on the proportions of hens perching, feeding and drinking, and the presence or absence of panting and wing spreading behaviors within each cage. The effects of the treatment, time period and treatment × time period interaction were considered. Perch use was higher at all time points in CP versus SP cages (P<0.0001) on the day of the heat stress event. Similar differences in perch use were not detected three days prior (P>0.05). Average mid-day perch use was higher on the day of the heat stress event (means: CP=58%; SP=45%) than three days prior (CP=20%; SP=18%). The onset of panting and wing spreading was delayed in CP cages as compared to SP and NP cages, and the incidences of these behaviors remained lower within CP cages as compared to SP or NP during and after the heat stress event (all P<0.05). This study provides preliminary evidence that the cooled perch system may assist laying hens in coping with heat stress. Further work is necessary to confirm the thermoregulatory benefits of the cooled perch system during prolonged heating episodes.

Environmental complexity and use of space in slow-growing free-range chickens under commercial conditions

Ane Rodriguez-Aurrekoetxea[1], Erin Hoerl Leone[2] and Inma Estevez[1,3]
[1]Neiker-Tecnalia, Animal Production, Arkaute Agrifood Campus, P.O. Box 46, 01080 Vitoria-Gasteiz, Spain, [2]Florida Fish and Wildlife Conservation Commission, FWRI, 1105 S.W. Williston Road, Gainesville, FL 32601, USA, [3]IKERBASQUE Basque foundation for Science, Alameda Urquijo 36, 48011 Bilbao, Spain; arodriguez@neiker.net

Nature provides a variety of tridimensional structures to animals that favors the development of complex behaviours. Contrarily, production environments for meat poultry generaly consist of bidimensional open areas where birds tend to cluster along the walls. We investigated the impact of increasing environmental complexity (EC) on slow-growing free-range chickens, raised under commercial conditions. The study was conducted in four farms, each consisting of three independent houses and outdoor ranges, with capacity for 1,300 birds/house. Chicks arrived on day one as a single flock (3,900) to each farm, remaining together in one of the houses until 5-6 weeks of age. At this age the flock was splited in. On each farm one house was outfitted with 9 panels indoors and 9 outdoors, one with 9 perches indoors and 9 outdoors, and the third house maintained standard management as a control. A total of 40 birds/house were tagged for individual recognition. Locations of tagged birds were recorded weekly (XY coordinates), from 5-12 weeks of age. Percentage of use of the central indoor areas, total (TD) and net (ND) distance travelled, and minimum convex polygons (MCP) were calculated, for indoor and outdoor areas. Production and welfare indicators were collected at the slaughterhouse and included; incidence of footpad dermatitis, fluctuating assymetry, and growth rate. Results indicated that the use of the central indoor area, while not affected by age ($P>0.05$), was affected by EC treatment ($P<0.05$), with greater use in houses with panels as compared to controls ($P<0.05$). A similar trend was found for the perch treatment ($P=0.057$). Indoors, the treatment affected TD ($P<0.05$), but no differences were found for TD outdoors ($P>0.05$). No treatment effect was found for ND, MCP, or the percentage of tagged birds using the outdoor range ($P>0.05$). However, both ND ($P<0.05$), TD ($P=0.054$), MCP outdoors ($P<0.05$) and the percentage of birds outdoors increased, or showed an increasing trend with age. No treatment effect was found for production and welfare indicators ($P>0.05$). The weak benefits of the EC treatments may have been restricted by the reduced number of devices introduced. Despite of this the results suggest that increased environmental complexity favours a more homogeneous use of the space inside the houses and that use of the outdoors increased with the experience provided by age.

Ramps in aviaries reduce falls and fractured keel bones in commercial laying hens

Ariane Stratmann[1], Ernst K.F. Fröhlich[2], Michael J. Toscano[1], Hanno Würbel[1] and Sabine Gebhardt-Henrich[1]
[1]*ZTHZ, Division of Animal Welfare, University of Bern, Burgerweg 22, 3052 Zollikofen, Switzerland,*
[2]*ZTHZ, FSVO, Burgerweg 22, 3052 Zollikofen, Switzerland; ariane.stratmann@vetsuisse.unibe.ch*

Housing laying hens in non-cage systems such as aviaries is believed to improve the welfare of the birds as aviaries provide more freedom to perform natural behaviour. However, these systems may increase the risk of falls and collisions with perches and housing structures, which is suggested to be one of the main causes of keel bone fractures. The aim of this study was to investigate how different aviary designs affect the prevalence of falls and collisions and whether these are related to the incidence of fractured keels. We compared three different aviary designs (including additional perches, platforms and ramps) with a control design (5 pens/design, 225 hens/pen). Video recordings of both pen sides were taken at 18, 23, 29, 37 and 43 weeks of age to assess numbers and reasons of falls (e.g. pushed by conspecifics, failed landing) as well as the occurrence of collisions. The keel bones of 100 hens per treatment (20 hens/pen) were palpated at the end of each video session to assess keel bone fractures. At 63 weeks of age all 400 focal hens were palpated, slaughtered and keel bones were collected for comparison with palpation data collected on the same birds. The experiment was carried out conforming to the legal requirements of Switzerland and all ethical guidelines were followed. Video data (video analysis focused on last 10 minutes of dusk and subsequent first 10 minutes of the dark period per pen side and age) were averaged per pen and age, palpation data were counted as absolute numbers per pen and age and both were analyzed using GLMM (RStudio 3.0.1) with pen as random factor. Reasons of falls (being pushed or failed landing) and collisions were both highly correlated with falls and thus excluded from the analysis to control for multicollinearity. Fewer falls occurred in pens equipped with platforms and ramps compared with control pens (platforms: $Z=-2.07$, $P=0.038$; ramps: $Z=-3.18$, $P=0.0015$). Also, fewer falls occurred at 37 and 43 weeks than at 18 weeks of age (37 weeks of age: $Z=-3.34$, $P=0.0009$; 43 weeks of age: $Z=-2.96$, $P=0.0031$). Fewer keel bone fractures occurred in pens equipped with ramps ($Z=-3.02$, $P=0.0026$). Incidence of keel bone fractures increased with age ($Z=9.02$, $P<0.0001$) but was not related to the number of falls ($Z=-0.51$, $P=0.61$). Results indicate that other factors than falls and collisions may cause fractures in laying hens in aviaries as well. However, reducing falls and fractured keels by including ramps in the aviary, is very likely to be beneficial for the birds and illustrates the impact housing design has on animal welfare and health.

The impact of rearing environment on nesting behaviour of laying hens in large furnished cages

Michelle E Hunniford, Stephanie Torrey, Gregoy Bédécarrats, Ian JH Duncan and Tina M Widowski

University of Guelph, Animal and Poultry Science, 50 Stone Rd E, Guelph ON N1G 2W1, Canada; mhunnifo@uoguelph.ca

An attractive nest site is considered to be one of the most important features of furnished cages for improving the welfare of laying hens. However, hens' perceptions of an appropriate nest site may be affected by their previous experience. A total of 1080 LSL-Lite hens were used; from day 1, half were reared in an aviary (access to shavings from 7 wks) and half in standard rearing cages. At wk 16, hens were housed in one of 2 sizes of cage [30 hens (n=12) and 60 (n=12) hens, both at 750 cm^2/hen] furnished with a nest area (red plastic curtains, yellow plastic mesh floor, 94 cm^2/hen) and a smooth red plastic scratch mat (50×50 cm) at opposite ends of the cage. Perches were parallel to the feeders in the centre of the cage. Daily egg location, determined by position on the egg belt, was recorded from 16-72 wks. Four hens from large cages and 2 from small were wing tagged at 18 wks and observed at the start of lay (wk 20) and at peak lay (~wk 32). Observers with handheld computer devices continuously recorded prelaying behaviour from lights-on at 0500 h until the hens oviposited. Frequencies and durations of behaviour observed during the 1 h prior to oviposition (n=106 observations) were transformed when necessary prior to analysis with Proc Mixed repeated measures (SAS 9.3). Because there were no significant interactions between cage size and rearing, only the effects of rearing are presented here (P>0.05). Cage-reared hens laid more eggs in the nest area (89.6±0.4% vs 81.3±0.7%; F=25.7, P<0.0001) and fewer in the scratch area (9.5±0.4% vs 18.0±0.7%; F=28.03, P<0.0001) than aviary-reared. Aviary-reared hens spent more time sitting on the scratch mat (13.8±2.9 min vs 11.2±2.9 min; F=4.01, P=0.048) but did not enter the scratch area more often than the cage-reared hens (5.6±0.9 vs 4.0±0.8; F=1.03, P=0.31). There was an interaction between age and rearing environment: at wk 20, cage-reared hens entered the nest area more often than aviary-reared hens (11.9±2.5 vs 5.1±1.2; t=-3.22, P=0.013) with no difference at wk 32 (5.4±1.7 vs 5.9±1.0; t=-0.61, P=0.92). Similarly, cage-reared hens were more active (walking + searching) than aviary-reared at wk 20 (17.2±2.4 min vs 11.3±2.0 min; t=-4.16, P=0.0009) but not at wk 32 (9.8±2.1 vs 11.4±1.8; t=-1.25, P=0.59). These results suggest that hens reared in an aviary found features of the nest less attractive than cage-reared hens but expressed more settled prelaying behaviour despite their laying more eggs in the scratch area.

Effects of weather conditions, early experience and vertical panels on slow-growing broilers' use of the free range area

Lisanne Stadig[1], Bart Ampe[1], Stefaan De Smet[2] and Frank Tuyttens[1]

[1]*Institute for Agricultural and Fisheries Research (ILVO), Animal Sciences Unit, Scheldeweg 68, 9090 Melle, Belgium,* [2]*Ghent University, Faculty of Bioscience Engineering, Coupure Links 653, 9000 Gent, Belgium; lisanne.stadig@ilvo.vlaanderen.be*

In chickens with access to a free range area, often only part of the animals go outdoors and they usually stay close to the barn. More frequent use of, and a more even distribution over the run can benefit animal welfare due to more space and opportunities to perform natural behaviours (e.g. foraging, dust bathing). The aims of this experiment were to investigate the effects of (1) weather conditions, (2) early experience with outdoor access and (3) vertical wooden panels in the outdoor area on free range use of slow-growing broilers. Sasso T451 chickens were housed in mobile stables starting from one week before they were allowed outdoor access (n=100 per stable, four stables in total). Each stable was placed centrally on a plot of 30×30 m, of which two quadrants consisted of open grassland and two of grassland with 30 vertical wooden panels (l×h: 2.5×0.6 m), placed parallel to each other. In each round (three in total), chickens from two stables were given outdoor access at 4 weeks of age, the other two stables at 5 weeks of age, both until 10 weeks of age. Ambient temperature, relative humidity (RH), precipitation, wind speed and solar radiation were recorded. Observations were performed two days per week, in the morning and evening, during which the number of chickens outdoor and their location were recorded. The effect of these factors on outdoor use and preference of outdoor area (with or without panels) was analysed using logistic mixed effects models with a random effect for stable to correct for the repeated measures. On average, 16.3% of the chickens were observed to be outdoors (range: 0-63.2%). Chickens used the free range area less with increasing wind speed (P<0.001), RH (P<0.001), precipitation (P<0.001) and solar radiation (P<0.001). The distance that chickens travelled from the stable increased with their age (P<0.001). This effect was more pronounced if access to the free range area was given at 4 weeks instead of at 5 weeks of age (P=0.042). Broilers had a preference for the outdoor quadrants with vertical panels as compared to the open grassland (56.2 vs. 43.8%; P<0.001). This preference was not significantly influenced by climatic parameters such as wind speed or precipitation. This could be explained by the absence of extreme weather conditions during the experiment, or it could mean panels provided insufficient protection against adverse weather. It can also indicate that chickens preferred the panels mainly because they provide a sense of safety.

Behaviour of free-range laying hens in distinct outdoor environments

Hannah Larsen[1], Greg Cronin[2], Paul Hemsworth[1], Carolynn Smith[3] and Jean-Loup Rault[1]
[1]Animal Welfare Science Centre, University of Melbourne, Parkville, 3010 Carlton, Australia, [2]University of Sydney, 425 Werombi Rd, 2570 NSW, Australia, [3]Macquarie University, Balaclava Rd, North Ryde, 2109 NSW, Australia; hlarsen@student.unimelb.edu.au

Free-range egg production systems provide hens with a variety of features in the outdoor range that are not provided indoor. However, the literature indicates that free-range laying hens do not consistently utilise the range in many commercial settings. In addition, the causation and function for the hens of utilising this outdoor range are unclear. We hypothesised that the behavioural time budget of hens will differ between distinct locations or 'patches' within the outdoor range, possibly explaining the observed pattern of spatial distribution across the range. The range characteristics were mapped on 1 commercial free-range farm with 8,000 62 week-old Hy-Line Brown hens and 4 distinct patch types were chosen on the basis of cover and substrate. Patch 1 (P1) consisted of 1 large Eucalyptus gum tree 30 m high and providing large canopy cover; patch 2 (P2) contained Acacia wattle trees providing 1-2 m high dense canopy cover; patch 3 (P3) consisted of bare sand and gravel ground; and patch 4 (P4) was bamboo-like dense leafy vegetation. All patch types were located 20 m from the shed and covered 38 m^2. Cameras captured 10 sec videos every 15 min over a 3 week period and 1 observer recorded behaviours using scan sampling. The number and behaviour of hens were recorded for each patch type and the data pooled per period of the day: 07:01-10:30 h, 10:31-14:00 h, 14:01-17:30 h, 17:31-21:00 h. Data were analysed with SAS using a mixed model. Patches 1 and 2 had higher numbers of hens than P3 and P4 ($P<0.05$). The interaction between location and period of the day ($P<0.05$) suggested that highly favoured areas are less subjected to diurnal patterns of range use whereas other areas saw more hens early or late in the day. The 5 most common behaviours were foraging, preening, locomotion, resting and vigilance. A wider variety of behaviours were observed in the highly preferred patches (P1 and P2) whereas mostly active behaviours (foraging and locomotion) were performed in areas that were less frequented (P3). These results support our hypothesis that different behaviours are performed in patches that differ in cover or substrate. Furthermore, providing areas of highly preferred types could enhance the spatial distribution across the range and minimise the diurnal pattern of range use in commercial settings.

Studies of tonic immobility duration in chickens

José Luis Campo, Sara García Dávila, María García Gil and María Teresa Prieto
Instituto Nacional Investigación Agraria y Alimentaria, Mejora Genética Animal, Carretera de
La Coruña km 7, 28040 Madrid, Spain; jlcampo@inia.es

The effect of several factors on the tonic immobility duration (a defensive anti-predator reaction and a behavioural measure of fearfulness in poultry) was studied in different chicken breeds housed in litter floor pens. The number of breeds was unequal in the different studies, ranging from 2 to 12. Tonic immobility was induced by placing the bird on its back with the head hanging in a U-shaped wooden craddle. Tonic immobility duration (from 0 to 600 s) was log transformed before analysis. A 2-way factorial analysis of variance of treatment (2 groups) and breed effects was used, after pooling pen effects with the residual. Experimental unit was the bird. The number of pens and birds was unequal in the different studies, ranging from 2 to 24 and from 2 to 20, respectively. Tonic immobility duration was significantly longer with moist litter (263 vs. 184±23 s; P<0.01), continuous light (429 vs. 243±7 s; P<0.001), noise (363 vs. 165±15 s; P<0.001), uncorrect collocation of the wing tag (260 vs. 184±32 s; P<0.05), supplemental allicin in heat stressed caged birds (215 vs. 90±16 s; P<0.01), pens vs. cages (221 vs. 152±17 s; P<0.05), vent pecking (294 vs. 172±17 s; P<0.001), footpad dermatitis (495 vs. 407±30 s; P<0.05), and rearing without a broody hen (445 vs. 308±24 s; P<0.001). The effect of auditory enrichment in caged birds was not consistent in all breeds (genotype-environment interaction was significant), the tonic immobility duration being significantly shorter only in one breed (141 vs. 330±19 s; P<0.05). There was no significant effect of stock density ranging from 20 to 4 birds per m^2 (214 to 230±26 s), physical enrichment (281 vs. 318±21 s), presence of perches (232 vs. 304±28 s), a droppings pit (249 vs. 261±27) or males (210 vs. 240±14 s), and outdoor vs. indoor housing (236 vs. 302±25 s), although housing effect approached levels of statistical significance (P<0.07). In conclusion, tonic immobility duration seems a consistent and reliable stress indicator, whereas access to environmental enrichment does not seem to have effect on fear behaviour.

Phenotypic appearance in laying hens: the looks matter

Irene Campderrich[1], Maria Guiomar Liste[1] and Inma Estevez[1,2]
[1]Neiker-Tecnalia, Arkaute Agrifood Campus, Department of Animal Production, P.O. Box 46, 01080 Vitoria-Gasteiz, Spain, [2]IKERBASQUE, Basque Foundation for Science, Alameda Urquijo 36-5 Plaza Bizkaia, 48011 Bilbao, Spain; icampderrich@neiker.net

Enriched cage systems offer greater behavioural opportunities to laying hens but larger flock sizes may also increase the risk of undesired behaviours. In addition, increased phenotype diversity of the group members (as expected in larger groups) may affect the way the group interacts. This study was designed to explore whether group size and the frequency of phenotypic diversity can affect the social dynamics in laying hens during the rearing phase. For the study, 1050 one day old Brown Hy-line chicks were randomly assigned to 45 pens at 3 group sizes (GS: 10, 20 and 40, constant density 8 pullets/m^2), combined with modified phenotypic appearance (PA) in different proportions; 0, 30, 50, 70 and 100%, by placing a black mark (non-toxic dye) in the back of the chicks head upon arrival. Therefore, populations were homogenous (100 or 0%), with all members of the group presenting the same marked (M) or unmarked (U) phenotype, or heterogeneous (30, 50, and 70%) where the two phenotypes (M and U) coexisted in the same pen. All pens were observed every other week from 3 to 13 weeks of age. We noted all aggressive and affiliative interactions across group members and their PA. As the frequency of interactions remained low, mean frequencies for the entire observation period were calculated per pen (to determine the effect of overall treatments) and type of bird giving and receiving the interaction (MM, MU, UU, UM, with the first letter representing the giver and the second the receiver). Data were analyzed with mixed model ANOVAs including GS and PA as fixed factors. GS, PA and type of interaction (MM, MU, UU, UM) were also used to determine the direction of the interaction in heterogeneous groups. Results showed that both aggressive and affiliative interactions occurred more often in small than large GS (P<0.05), regardless of the PA treatment (P>0.05); GS × PA (P>0.05). However, in heterogeneous groups (30, 50 and 70%), higher than expected (standardized by frequency of the specific phenotype) incidence of total aggressive interactions were observed in GS 20 from U towards M (P<0.05) as compared to all other phenotype combinations. On the contrary, less than expected aggressive interactions were found between MM birds in GS10. The results suggest that both aggressive and affiliative interactions may be determined by the opportunities to interact and that the negative effects of phenotypic diversity are more apparent in small and medium GS as compared to large GS.

Nature of agonistic behaviours between sows mixed at different space allowances

Jean-Loup Rault and Rachel Taylor

Animal Welfare Science Centre, University of Melbourne, Parkville, 3010 Carlton, Australia;
raultj@unimelb.edu.au

In confined housing systems, the limited space provided to animals should be sufficient to enable the display of adequate social behaviours. The inability to escape aggression in particular is a welfare concern. Most studies report 'aggression' as a whole without dissociating between different types of aggressive behaviours. We hypothesised that the type of social behaviours between group-housed sows is affected by space allowances (1.45-just above minimum industry standards of 1.4, 2 or 2.9 m^2) or day after mixing (day 2 vs. 9). Sows were mixed in unfamiliar groups within 5 days of insemination at 1.45, 2 or 2.9 m^2 per sow by placing 20, 14 or 10 sows per pen. For each space allowance 5 pens were studied with 10 focal sows per pen, for a total of 150 sows. Social behaviours between focal sows were recorded for 15 min after drop-feeding the day after mixing and 1 week later using video and an ethogram of 19 behaviours. Blood samples were collected from 3 sows per pen on day 2 and 26 for analysis of cortisol and progesterone concentrations. Data were analysed in SAS using Chi-square or the Kruskal-Wallis tests since data were not normally distributed. This project was approved by the Rivalea Animal Ethics Committee. A total of 587 interactions were observed, with 60% being brief interactions composed of 1 reciprocal action. The number of interactions was not influenced by Space or Day (both P=0.08). The mean duration of interaction was not influenced by Space (P=0.11) but decreased from day 2 to day 9 (P=0.005). The aggressor displayed more bites and headpushes/headknocks at smaller spaces (P=0.05 and P=0.005) whereas threats (attempt to bite or headknock) and socio-neutral behaviours (nose interactions) occurred more at larger spaces (P=0.01 and P=0.04). Bites and socio-neutral behaviours reduced from day 2 to 9 (P=0.02 and P<0.0001). The recipient did not modify their behaviour according to Space or Day apart from a higher occurrence of no reaction on day 2 compared to day 9 (P<0.0001). The occurrences of retreat, chase or a third sow getting involved were not affected by Space or Day. In 90% of cases, the interaction was terminated by the recipient who retreated (77%) or showed no reaction (10%) but this was not influenced by Space. Cortisol decreased and progesterone increased from day 2 to 26 (both P<0.01) but were not affected by Space, although this is likely due to the low sample size for cortisol. Space was confounded with group size in this study, but the literature indicates that group size has a negligible effect in comparison to space for sows. This detailed study shows that sows use different social strategy depending on the space available, with more severe aggressive behaviours in smaller spaces. These differences disappeared by day 9.

Stimulating exploratory behaviour in piglets: effects on pre-weaning creep consumption

Yolande M Seddon, Krista Davis, Megan Bouvier and Jennifer A Brown
Prairie Swine Centre, Ethology and Welfare, Box 21057, 2105 8th Street East, Saskatoon, S7H 5N9, SK, Canada; yolande.seddon@usask.ca

This work investigated whether pre-weaning creep consumption can be increased through stimulating exploratory behaviour in piglets, and whether this is best achieved through provision of enrichment (E) to increase pen exploration, or rather through presentation of creep in a large shallow feeder to facilitate synchronized feeding among littermates. Four treatments, (n=7 litters/treatment) were studied: T1: Creep provided in a standard feeder (SF); T2: creep provided in SF, with E provided; T3: creep provided in a large tray feeder (TF); T4: Creep provided in TF with E provided. Creep feed was offered to all litters from 10 days after birth until weaning (28 days). Strips of cotton rope were made available to litters in the E treatment from 5 days after birth until weaning. Piglet weights and creep consumption/pen were recorded weekly. Behaviour was recorded on four litters per treatment for 8 hours (8 am to 4 pm), on days 12, 19, and 26. Footage was scanned at five minute intervals and the number of piglets interacting with the feeders (head in feeder) recorded, and for E treatments, the number of piglets interacting with E (snout touching, or chewing E). The frequency and the average number of piglets per visit observed at the feeder or E was calculated on each observation day. Pen gain and creep consumption were calculated. Data were analysed using PROC MIXED (SAS 9.2) to determine the individual and interactive effects of feeder type and provision of E on visits to the creep. Piglets provided with E were seen to contact the E on average 11 times per day. Feeder type, but not E, resulted in a greater frequency of piglet visits to the feeder on day 12 (SF: 1.25, TF: 6.0, SEM 1.22, $P<0.05$), with a tendency for a greater average number of piglets at the TF per visit ($P=0.060$). At day 26 there was a tendency for a greater frequency of visits to the TF (SF: 5.33, TF: 16.38, SEM 4.05, $P=0.086$). Litters supplied with TF had a greater daily creep disappearance (SF: 86.3 g, TF: 163.4 g, SEM 20.4, $P<0.05$), with no effect of E. However, litters provided with the SF had a greater piglet birth to wean average daily gain (SF: 0.25 kg, TF: 0.22 kg, SEM 0.01, $P<0.05$), with no effect of E. Provision of a larger feeder that encourages social feeding, rather than E to encourage exploratory behaviour, appears to have a greater influence on attracting piglets to creep feed. The increased creep disappearance found in the TF suggests piglets were interacting with the creep. However, birth to wean growth rates did not reflect any benefits of increased creep consumption suggesting that use of TF may have resulted in greater feed wastage.

Housing related injuries in Swedish horses: causes and preventative actions

Jenny Yngvesson[1], Anna Stigwall[1], Carolina Leijonflycht[1] and Michael Ventorp[2]
[1]Swedish University of Agricultural Sciences, Dept of Animal Environment & Health, P.O. Box 234, 532 23 Skara, Sweden, [2]Swedish University of Agricultural Sciences, Dept of Biosystems & Technology, P.O. Box 52, 23053 Alnarp, Sweden; jenny.yngvesson@slu.se

Horses are animals with a rapid and powerful flight response. This has consequences for both horse and human safety. This study aimed to investigate what housing factors pose risk of injury to horses in Sweden, with a total population of around 350,000 horses. Two partial studies were carried out during the autumn 2013. Firstly, the regional veterinary stations were contacted and interviewed about horses with injuries they had treated, the causes of the injuries and the severity of the injuries. Secondly, horse owners with horses injured by building structures, fittings and equipments, were recruited through social media and interviewed about causes of the injuries and how the treatment efficacy. Four of the 30 veterinary stations answered our questions. During the last two years they had treated 15 horses with housing related injuries. Of these, nine were injured on pasture or in paddocks, four in the stable, one in a horse walker and one in a single horse trailer. Most commonly, the cause was either unknown, fence or box furnishing. Seven of the injuries were severe, of which three were euthanized, four were moderate and four were mild. Horse owner interviews identified 24 horses with housing related injuries. Of these, 20 were injured on pasture or in paddocks, three in stables and one in a horse walker. Of the ones injured on pasture six were severe and five were euthanized. The most common cause of injury was a fence and the second most common was another horse. These results show that our housing of horses does not always comply with their ethological needs of social contact with other horses and freedom of movement. We conclude that to reduce housing related injuries in horses' fences should be well designed and in good repair to minimize horses getting caught in wire and electric bands. Further research is needed and horse owner knowledge about how to introduce unfamiliar horses to each other needs to be increased.

Effect of full enrichment on behaviour and welfare of lambs

Lorena Aguayo-Ulloa[1], María Pascual-Alonso[1], Morris Villarroel[2], Genaro Miranda-De La Lama[3] and Gustavo María[1]
[1]*University of Zaragoza, Faculty of Veterinary, Animal Production and Food Science, Miguel Servet 177, 50013 Zaragoza, Spain,* [2]*Polytechnic University of Madrid, Department of Animal Science, Ramiro de Maeztu, 7, 28040 Madrid, Spain,* [3]*Metropolitan Autonomous University, Food Science, Av. de las Garzas 10, 52005 Lerma, Mexico; laguayo@unizar.es*

Sixty male lambs (65 days old, 17.13±0.18 kg), were housed for 5 weeks in 6 pens (2.9×3.3 m, 0.95 m²/lamb, 10 lambs/each) in 2 treatments (3 replicates/treatment). The conventional system (CO) had no enrichment. The enriched system (EE) had straw (forage and bedding), a feeder ramp (FRA) and a play ramp (PLP). Behaviour, use of spaces and social interactions were recorded daily on week 4 (scan sampling, 8:00 to 20:00, 1'/10') and continuous sampling (7 h/day). During week 5, lambs were subjected to a T-maze on 2 consecutive days. Blood samples and infra-red temperatures (IRT) were taken to assess stress response. Behaviour data were transformed by the SQRT function. The social behaviour variables were analysed using PROC MIXED with repeated measurements. T-maze variables were subjected to repeated-measures ANOVA. Stress response variables were analysed by GLM of SAS. Lambs spent most of their time resting (78%) and EE rested less than CO lambs in the morning ($P<0.05$). Walking and feeding on concentrate were higher in the morning ($P<0.05$). EE lambs spent more time foraging in the morning (6%) and stood up more in the morning (11%) than CO lambs ($P<0.05$). EE had higher ($P<0.05$) agonistic and affiliative interactions than CO lambs. Stereotypes were higher in CO ($P<0.05$). EE lambs solved the T-maze more quickly and crossing less areas than CO ($P<0.05$). Overall, lambs resolved the second exposure faster than the first ($P<0.05$). Glucose levels were higher in EE lambs, and NEFA ($P<0.05$) and the neutrophil/lymphocyte (N/L) ratio ($P<0.05$) were higher in CO, indicating chronic stress probably related to a low level of stimulation. EE lambs had better adaptation process to the novel environment, decreasing their biological cost and optimizing their welfare. The social stability reach by lambs at the end of the fattening period was not affected by the enrichments.

Spacing behavior of cows in free-stall housing containing different breeds in a group

Shigeru Morita, Mai Honda, Michiko Ueda and Shinji Hoshiba
Rakuno Gakuen University, Ebetst, 069-8501 Hokkaido, Japan; smorita@rakuno.ac.jp

The objective of this study was to investigate the spacing behavior of cows in free-stall housing when cows lay down in the stall. Thirty-one observations were made for individual stall selection in a free-stall housing that kept Jersey cows (average 5 cows) and Holstein cows (average 35 cows) together. The observation was done once a day, at approximately 07:00 in the morning. Cows were milked twice (05:30 and 16:30) and offered TMR (1000) once a day. The average body weight of Jersey cows was 482 kg and of Holstein cows was 657 kg. The number of stalls was 42. The stalls were placed in two rows, with cows facing each other. The neighboring cows were defined as cows lying on the facing or side stalls. For illustration of the connection of neighboring cows and analysis of the index of network graph, Pajek (ver. 2.05) was used. The Kruskal-Wallis test was used for comparing averages. The number of cows in stalls at observation time were from 16 cows to 38 cows (14-33 Holstein cows and 2-5 Jersey cows). The stall placed in the end of stall row was chosen less than other stall in both breeds (0.5% in the end of stall row vs 2.5% in the other position in Jersey, 0.8% vs 2.5% in Holstein). Jersey's choice for individual stall was varied wider (from 0.0 to 8.8%) than Holstein's choice (from 0.3 to 3.3%). The average distance between two cows was 6.0 m for Jersey cows, 8.2 m for Holstein cows, and 7.8 m for different breeds. The distance between Jersey cows was significantly ($P<0.05$) shorter than that between other pairs of cows. The frequency of individual neighboring of cows ranged from 0 to 6 times. In the network of neighboring cows that is equal and greater than two times, 48 cows linked to one another, but only one cow did not link with other cows. The density of network was 0.15, average degree was 7.3, diameter was 5 and average path length was 2.3. The average degree of Jersey cows (10.5) tended ($P=0.08$) to be more than that of Holstein cows (7.0). The average path length between Jersey cows in the network graph was 1.33, and this was significantly ($P<0.05$) shorter than that for Holstein cows, and as well as for different breeds. Jersey cows connected within two lines of other Jersey cows in the network graph. There were no difference between Jersey and Holstein cows for the centrality of closeness and betweeness. It was concluded that Jersey cows had more neighboring cows than Holstein cows and shorter path length in network relationship when two breed were kept in a group.

Behaviour of lactating cows under automatic milking system: a case study

Keelin O'driscoll[1], Dayane Teixeira[1], Daniel Enriquez-Hidalgo[2], Bernadette O'brien[2] and Laura Boyle[1]
[1]Teagasc, Pig Development Department, Moorepark, Fermoy, Co. Cork, Ireland, [2]Teagasc, Livestock Systems Department, Moorepark, Fermoy, Co. Cork, Ireland; keelin.odriscoll@teagasc.ie

There is growing interest in the use of automatic milking systems (AMS) to reduce the time cows spend away from pasture. However, little is known about cow behavior in pasture based systems. This study describes patterns of behavior in pasture based cows milked by AMS. Cows voluntarily presented themselves to the AMS for milking. The land area was 24 ha divided in three grazing sections. Three roadways radiated from the AMS. Water was only available in the AMS holding yard. From a herd of 63 cows, 22 spring calving (calving: 18 Jan - 12 Apr) Holstein-Friesian (n=19) and Norwegian Red (n=3) multiparous (parity 2-5) cows with an average milk yield of 11.7±3.24 l/cow/d on the week before start of the experiment were used in the study. Position (AMS/pasture/roadway), posture (lying/standing) and behavior (grazing/ruminating/idling/other) were recorded on 4 days from 06:30 to 18:45 and on 1 day from 06:30 to 16:45 between Aug and Sept 2011, using scan sampling at 10 min intervals. Standing/lying was recorded every 30 sec for 3×24 h periods using modified voltage dataloggers (Tinytag Plus, Chichester, UK). The AMS recorded the time of milking events. Associations between variables were determined using Spearman correlations and regression analysis (CORR and Mixed procedures, SAS v9.3). Cows had 1.8±0.5 visits to the AMS per day, and the average number of visits peaked in the hours prior to 02:00, 09:00 and 16:00. The proportion of time spent lying per hour was highest between 03:00 (78%) and 06:00 (88%). The average time between standing and a milking event was 01:15±01:11. Cows spent 79±5.1% of the observation time at pasture, 14±3.9% in the AMS area and 7±2.7% on roadways. Cows were mainly idle in the AMS area (43±12.2%) and grazing while at pasture (55±10.0%). Cows lay for 26±8.0% of visual observation time; 99±11.4% of lying observations were at pasture, and 1±9.7% at the roadways. Cows spent 10:35±01:42 (hh:mm; mean ± s.d.) lying per day, in 7.21±2.13 lying bouts with an average duration of 01:19±00:57 hh:mm. Date had an effect on daily lying time (R^2=0.259; P<0.001) and the number of lying bouts per day (R^2=0.300; P<0.001). There was a negative correlation between DIM and lying behavior (r=-0.437; P=0.04) and a tendency for a negative correlation between DIM and time at pasture (r=-0.391; P=0.07). The data suggest that cows in an AMS have a different daily pattern of behavior to 2×daily milked cows. Thus AMS systems could improve dairy cow welfare by permitting cows a greater level of control over their daily time budgets than traditional systems.

Seasonal effect of white clover inclusion in grass swards rotationally grazed on dairy cow feeding behaviour

Daniel Enriquez-Hidalgo[1,2], Trevor Gilliland[1,3] and Deirdre Hennessy[2]
[1]*Queen's University Belfast, Institute for Global Food Security, School of Biological Sciences, BT7 1NN, Belfast, United Kingdom,* [2]*Teagasc – Moorepark, Animal & Grassland Research and Innovation Centre, Fermoy, Co. Cork, Ireland,* [3]*Agri-Food and Biosciences Institute, Plant Testing Station, Crossnacreevy, BT5 7QJ, Belfast, United Kingdom; daniel.enriquez@teagasc.ie*

Cows have a preference for white clover (WC) compared to grass, but this preference may be affected when restrictions are imposed, which can occur in rotational grazing systems. This experiment compared the seasonal effect of WC inclusion to grass swards with grass only (GO) swards rotationally grazed on dairy cow feeding behaviour. Forty cows were randomly allocated to graze each sward. Cows received 17 kg DM/cow/d. Herbage mass and WC content were estimated twice weekly. Feeding behaviour was estimated over a two-week period using 28 cows in late-spring and 36 cows in summer and autumn. Each cow was fitted with a grazing behaviour recorder for two consecutive days in each period. Data were analysed using PROC MIXED in SAS. Herbage mass and WC content were 1.51 t DM/ha and 17%, respectively, in late-spring; 1.70 ton DM/ha and 30%, respectively, in summer; and 1.24 ton DM/ha and 31%, respectively, in autumn. In spring, cows spent 10% more time grazing when WC was present (608 and 551±15.6 min/d; P<0.05), but WC had no effect on ruminating or idling times (418±12.7 and 459±20.3 min/d, respectively). The difference in time grazing was mainly driving by longer night grazing time (17.1 and 13.8±0.84 min/h; P<0.01) at the expense of less night idling time (24.4 and 21.9±0.76 min/h; P<0.01). In summer, cows had similar grazing and idling times (590±11.1 and 390±13.4 min/d), but spent 8% less time ruminating when WC was present (443±11.2 and 480±13.8 min/d; P<0.05), with greatest differences at night (30.9±0.49 and 33.1±0.59 min/h; P<0.01). Similar daily grazing (594±10.1 min/d), idling (398±18.4 min/d) and ruminating (448±13.5 min/d) times were observed in autumn, but cows spent 5% less time ruminating at night when WC was present (30.3 and 31.9±0.59 min/h; P<0.05). Cows may have been motivated to search for WC in spring due to its low content in the sward and probably as WC was concentrated in patches. In summer and autumn there was more WC, probably more homogeneously distributed, reducing foraging requirement. WC reduced rumination time when it's content was greatest in summer, but only at night time in autumn, probably related to the lower herbage mass that both swards had in autumn. The WC inclusion into GO swards had an effect on cow feeding behaviour closely related to WC content; grazing time increased when WC content was low, due to an increase in foraging behaviour, and ruminating time was reduced when WC content was greater.

Preference for rearing substrate of dairy calves
Mhairi A. Sutherland, Gemma M. Worth, Mairi Stewart and Karin E. Schütz
AgResearch Ltd, Ruakura Research Centre, Hamilton, New Zealand;
mhairi.sutherland@agresearch.co.nz

Rearing substrate is an important component of the pre-weaning environment of dairy calves. Traditional substrates, such as sawdust, can be difficult and/or expensive for farmers to obtain in New Zealand, therefore there is a need to evaluate alternative rearing substrates for dairy calves that that are economically viable, readily available and support an acceptable level of animal welfare. The aim of this study was to investigate the preference of dairy calves for four different rearing substrates. At 1 wk of age, 24 calves were housed in groups of four in pens which were evenly divided into four rearing substrates: sawdust (SD), rubber chip (RC, 4-7 mm in diameter) created from recycled car tyres, washed river sand (SA) and quarried stones (ST, 20-40 mm in diameter). Calves were reared on SD prior to the start of the experiment. During the first 3 days calves were given free access to all four substrates. Calves were then restricted to each substrate type for 48 h. In order to rank preference, calves were then exposed to all surfaces in a pair-wise manner for 48 h. All calves experienced all six combinations of substrate pairs. Finally, calves were again given free access to all four substrates simultaneously for 48 h. Lying behaviour and position in the pen were recorded for 24 h at the end of each experimental period using camcorders and accelerometers (Onset Pendant G data loggers). Preference was determined based on lying times on each substrate. Data were analysed by ANOVA. Calves preferred lying on SD than the other 3 surfaces (SD versus RC: 86% time spent lying on SD; SD versus SA: 92% time spent lying on SD; SD versus ST: 93% time spent lying on SD, P<0.001). Calves preferred lying on RC than SA and ST (RC versus SA: 86% time spent lying on RC; RC versus ST: 87% time spent lying on RC, P<0.001). Lastly, calves preferred lying on SA than ST (SA versus ST: 86% time spent lying on SA, P=0.001). During the initial free access period, calves spent more (P=0.015) time lying on SD than all other substrates (SD: 86±7.8%, RC: 5±6.1%, SA: 3±3.9% and ST: 3±2.9% total observations/24 h, means±SD). At the end of the study when given free access to all rearing substrates, calves spent more (P<0.001) time lying on SD than all other substrates (SD: 84±6.3%, RC: 5±2.7%, SA: 7±4.0% and ST: 4±1.6% total observations/24 h, means±SD). Dairy calves showed a clear preference for sawdust over other rearing substrates and the least preference for stones. These results suggest that a rearing substrate that is compressible with good insulation properties is important to dairy calves.

Summertime use of natural versus artificial shelter by cattle in nature reserves

Eva Van Laer[1], Bart Ampe[1], Christel Moons[2], Bart Sonck[1], Jürgen Vangeyte[3] and Frank Tuyttens[1,2]
[1]*ILVO (Institute for Agricultural and Fisheries Research), Animal Sciences Unit, Scheldeweg 68, 9090 Melle, Belgium,* [2]*Ghent University, Faculty of Veterinary Sciences, Heidestraat 19, 9820 Merelbeke, Belgium,* [3]*ILVO (Institute for Agricultural and Fisheries Research), Technology and Food Science Unit, Burgemeester Van Gansberghelaan 115 bus 1, 9820 Merelbeke, Belgium; eva.vanlaer@ilvo.vlaanderen.be*

Cattle grazing nature reserves may encounter aversive weather conditions, including heat-load in summer. Whether artificial shelter (AS) should be provided in addition to natural shelter (NS; vegetation) is debated. We aim to determine how climatic factors affect cattle's summertime use of AS vs. NS. In eight reserves with an AS and varying vegetation cover, GPS-collars monitored 24 h terrain-use of one animal during one (n=2) or two (n=6) summers. We plotted animal positions on digital maps to determine when cows used open area (OA), NS and AS (including 5 m around them). Weather-stations provided open-field climatic measurements and calculations of heat-stress indices, for example, the Heat Load Index (HLI). The general preferences and circadian patterns in use of OA, NS and AS were graphed. The use of OA was related to wind-direction by mixed-model ANOVA and to other climatic variables by mixed logistic regressions. In most reserves cattle used less OA and more NS during the daytime. In each reserve, AS covered <0.5% of the area. In six reserves it was also used <0.5% of the time (mean±SE = 0.10±0.05%). In the most sparsely vegetated and another – the most vegetated – reserve, AS was used 9.5% and 2.5% of the time, respectively (both mainly during daytime). Use of OA was significantly but differently influenced by wind-direction in seven reserves, decreased with increasing wind-speed (P<0.01, n=5) and rain-intensity (P≤0.01, n=4) in most reserves, and always decreased with increasing HLI (P<0.001). Additional mixed logistic regressions show that in the most vegetated reserve increasing HLI increases the use of NS vs. AS steeply vs. minimally, respectively (P<0.0001). In the least vegetated reserve increasing HLI decreases vs. increases the use of NS vs. AS (P<0.0001). These findings suggest that sheltering behaviour of cattle is reserve-dependent, but influenced by heat-load in all studied reserves. If sufficient NS is available, cattle probably rather use this than AS for protection. If little NS is available cattle use AS. The need for additional protection against heat-load for cattle in nature reserves thus seems to depend on the availability of NS.

Space allowance for confined livestock: minimum legislative limits, allometric principles and best practice compared

Tracey Jones

Compassion in World Farming, River Court, Mill Lane, Godalming, GU7 1EZ, United Kingdom; tracey.jones@ciwf.org

The space we provide our farmed animals determines their freedom of movement, whilst the quality of the space provided determines the behaviours they are able to express. Both movement and behavioural expression are integral to wellbeing, yet neither is considered when setting minimum space allowance limits. Space provision tends to be limited to the static requirement of an animal to stand and lie down, and is influenced in practice by the animal's thermal requirements, ventilation capacity of the house, and floor type. In general, housing with forced ventilation systems, barren environments and fully slatted floors, provide the least space, so least opportunity for movement and behavioural expression. Limiting space has negative consequences on production usually beyond the point at which behaviour is restricted. The amount of space an animal occupies as a consequence of it shape and size can be estimated using the allometric equation $A=kW^{0.67}$ where A is given in m^2/animal, W is body weight in kg and k is an empirical constant. Space requirement therefore increases non-linearly as body weight increases, and increases as the value of k increases. The k values for pigs to stand with negligible space for movement, to lie in sternal recumbency, or to lie in lateral recumbency are 0.019, 0.037, and 0.0457, respectively. Legislative minimum space allowance limits are based on allometric principles for pigs, albeit in a stepped increments as opposed to a smooth curve, and represent the space required for pigs in sternal recumbency (example 0.65 m^2 for a 100 kg pig). The space required to reduce the risk of tail biting in pigs is however 1 m^2/100 kg pig; the calculated space on the lateral curve for a 100 kg animal. For other species, legislative limits or recommendations take a flat line approach to stocking density across a range of final body weights. The point at which the flat line is taken according to k value and body weight are given for broiler chickens (33 kg/m^2 is the point at which a 1.8 kg bird hits the sternal curve for space allowance), and laying hens (9 hens/m^2 is the point at which a 2.1 kg bird hits an extrapolated curve with a k value of 0.068). The first extrapolated k value is taken from the FAWC recommendation of 25 kg/m^2 for a 5 kg turkey. Allometric principles for pigs and poultry, indicating space allowances for sternal, lateral, and two extrapolated k values are presented. The impact of adopting a flat line approach to stocking density on lower and higher weight animals and raising the k value according to activity levels are discussed. Finally, an allometric approach to studying the effect of different space allowance / stocking density on animal welfare is recommended.

Chronic positive experience with humans induces optimistic judgement bias in weaned piglets

Sophie Brajon[1,2], Océane Schmitt[2], Jean-Paul Laforest[1] and Nicolas Devillers[2]
[1]*Université Laval, 2325 rue de l'Université, Québec, G1V 0A6, Canada,* [2]*Dairy and Swine R & D Centre, 2000 rue college, Sherbrooke, J1M 0C8, Canada; sophie.brajon@agr.gc.ca*

This study investigates whether a positive or a negative chronic experience with humans can induce judgement bias by modulating emotional state in piglets. After weaning, 29 piglets were subjected to a repeated experience with humans: gentle contacts (n=8, GEN), rough handling (n=10, ROU) or minimal contacts (n=11, MIN). Simultaneously, they were individually trained to discriminate a positive auditory cue (P) associated with a reward (cereals) and a negative one (N) associated with punishments (ex: air spray) and delivered into a test box containing a trough. Piglets received two sessions of six 40-sec trials per day for at least 9 days. When piglets successfully discriminated both cues (approach following P or do not approach following N), they were subjected to cognitive bias tests (CBT) including a new ambiguous auditory cue (A, 50/50 between P and N). The A cue was neither rewarded nor punished. CBT were performed in the presence or absence of the human who delivered the treatment. The approach rate following A cue was corrected from the average individual response to P and N cues $(=(X_{app\_P} - app\_A)/(X_{app\_P} - X_{app\_N}))$. A low corrected approach is associated with a high approach rate following A cue compared with the average individual response whose intermediate value would be 0.5. The average approach rate and corrected approach were analysed using mixed models and the latency to approach curve was modelised using non-parametric Kaplan-Meier method. During CBT, all piglets discriminated between P and N cues (approach rate, P: 81±3%, N: 27±3%, P<0.001). As expected, corrected approach following A cue was lower for GEN piglets than ROU piglets (GEN: 0.42±0.28, ROU: 1.59±0.31, P=0.04). MIN piglets had intermediate but not significantly different values (MIN: 0.73±0.33). Latency to approach curves were compared and a significant difference was observed with A cue for GEN compared to ROU piglets (P=0.02), where GEN piglets approached the through more quickly than ROU piglets. Presence of the human when performing CBT did not affect piglets corrected approach following A cue (with H: 0.74±0.09, without H: 0.94±0.09, P=0.34) nor latency to approach (P=0.41) for all treatments. This study is the first one to demonstrate a cognitive bias in weaned piglets in relation to the valence of previous experience with humans. Piglets that received a chronic positive experience with humans showed a slight optimistic judgement bias compared to piglets receiving a chronic negative experience which is indicative of a more positive affective state. However, the presence of the human during CBT did not influence the piglets' responses.

Pre-weaning iron deficiency in piglets impairs spatial learning and memory in the cognitive holeboard task

Alexandra Antonides[1,2], Anne C. Schoonderwoerd[1,2], Brian M. Berg[3], Rebecca E. Nordquist[1,2] and Franz Josef Van Der Staay[1,2]
[1]Rudolf Magnus Institute of Neuroscience, Universiteitsweg 100, 3584 CG Utrecht, the Netherlands, [2]Emotion and Cognition group, Department of Farm Animal Health, Faculty of Veterinary Medicine, Yalelaan 7, 3584 CL Utrecht, the Netherlands, [3]Mead Johnson Pediatric Nutrition Institute, 2400 West Lloyd Expressway, Evansville, IN 47712, USA; a.antonides@uu.nl

Iron deficiency is the most common nutritional deficiency in humans, affecting more than two billion people worldwide. Iron is an essential nutrient, required for many biological functions. Iron deficiency in early life can lead to irreversible deficits in learning and memory. The pig represents a promising model animal for studying these deficits, because of similarities to humans in early development. This study investigated long-term effects of pre-weaning iron deficiency on cognitive performance in a spatial cognitive holeboard task using piglets. Ten piglets were fed an iron deficient milk diet (21 mg iron/kg diet) and ten siblings a control milk diet (88 mg/kg) for four weeks. Then, all piglets were fed a fully balanced commercial pig diet (190-240 mg iron/kg). Piglets were then trained and tested in the cognitive holeboard task in which 4 of 16 holes contained a food reward, which allows measuring working memory (ratio of rewarded site visits divided by the total number of visits to rewarded set of holes; short-term memory) and reference memory (ratio of visits to the rewarded set of holes divided by the total number of visits; long-term memory) in parallel. Piglets were trained for 40-60 trials during the acquisition phase, then transitioned to the reversal phase, in which a different set of four holes was baited. A mixed model repeated measures ANOVA (SAS PROC MIXED) revealed that iron deficient piglets showed an impaired reference memory performance during acquisition ($F=8.66$; $df=1,14$; $P=0.011$) and the transition phase ($F=4.83$, $df=1,14$; $P=0.045$). These results confirm the hypothesis that early iron deficiency leads to lasting cognitive deficits. The holeboard test and the piglet as model animal can be used in future research to assess long-term cognitive effects of diets fed before weaning.

91-days-old piglets recognize and remember an old aversive handling

Roberta Sommavilla[1], Evaldo Antonio Lencioni Titto[1], Maria José Hotzel[2], Cristiane Gonçalves Titto[1], Lina Fernanda Pulido Rodrigues[1], Thays Mayra Cunha Leme[1], Alyne Suesique Sampaio[1] and Alfredo Manuel Franco Pereira[3]
[1]University of São Paulo, Duque de Caxias, 225, Pirassununga, SP, 13.635-950, Brazil, [2]Federal University of Santa Catarina, Rodovia Admar Gonzaga, 1346 Itacorubi, Florianópolis, SC, 88034-001, Brazil, [3]University of Evora, Largo dos Colegiais 2, Évora, 7004-516, Portugal; robertasommavilla@gmail.com

The aim of this study was to perceive if piglets can recognize their aversive handling and if they can remember this aversive handling after three weeks with no contact with this person. For this, 16 piglets received an aversive treatment during their first 70 days of life: between day 1 and 28 this treatment was applied daily and, from day 29 to day 70, during alternate days. This aversive treatment was made by the same person (AH), a woman wearing orange coveralls and black boots. For the aversive treatment, AH was noisy, moved harshly and unpredictably and shouted frequently during routine cleaning of facilities and animal handling. After day 70 they received a gentle treatment with another person (GH – a woman wearing dark blue coveralls and white boots) and never had contact with AH again. For the gentle treatment, GH used a soft tone of voice and was careful during the same routine. The Human Approach Test was applied to measure the avoidance response of piglets to the approach of AH and an unfamiliar handler (UH) in a novel place, at 35 days and at 91 days after birth. Scores ranged from 1 (experimenter could touch piglets) to 5 (piglets escaped as soon as person moved). The UH was wearing white coveralls and white boots. Data were analyzed by t-test at day 35 and ANOVA and Tukey at day 91. At 35 days, piglets kept more distance from AH than from the UH (2.37 ± 0.33 and 1.69 ± 0.22 respectively, $P=0.04$), indicating that the piglets could recognize the aversive handler. At 91 days, piglets still kept more distance from AH than UH and GH (2.75 ± 0.33; 1.31 ± 0.15; $0.25\pm0,11$ respectively, $P<0.001$), indicating that aversively treated piglets do not avoid an unknown handler, but still can remember the aversive handler with whom they had contact early in life. In conclusion, 35 and 91 days-old piglets show different avoidance responses to a human, according to the quality of handling received. Moreover, they recognize their handler and remember her after three weeks with no contact.

Effect of the human animal relationship on cognitive bias test in pigs

Ricard Carreras[1], Eva Mainau[1,2], Antoni Dalmau[1], Xavier Manteca[2] and Antonio Velarde[1]
[1]IRTA, Animal Welfare Subprogram, Veïnat de Sies, s/n, 17121 Monells (Girona), Spain, [2]Universitat Autònoma de Barcelona, Animal and Food Science Department, Veterinary Faculty, 08193 Bellaterra (Barcelona), Spain; ricard.carreras@irta.cat

The aim of this study was to assess the effect of the human animal relationship (HAR) on cognitive bias (CB) in pigs. Two CB tests were performed, one before (pre-treatment CB test) and the other after (post-treatment CB test) the HAR treatment. A total of 48 female pigs of 11 weeks of age coming from the same commercial farm were individually trained during 11 sessions to discriminate between a rewarded (R) and unrewarded (U) bucket in a test pen according to its position. In each training session the bucket was allocated in the left or right side of the pen. R bucket was situated in the left side of the pen for half of the pigs and in the right side for the other half. Afterwards, each animal was subjected individually to the pre-treatment CB test, where the bucket was placed in a central position (no left, no right). Both training sessions and CB test finished 30 seconds after the pig ate or tried to eat apples or 5 minutes after the pig entered the pen. The time to contact the bucket was recorded in all sessions and CB test. After the pre-treatment CB test, pigs were allocated into 2 different rooms (GH and BH), with 4 pens of 6 pigs each. Pigs in the GH room had a 'good' HAR as they were handled gently while pigs in the BH room had a 'bad' HAR, as they were exposed to sudden noises when the handler was close to the animal and immobilized individually during periods of 20 seconds. The HAR was performed during 5 minutes per pen and day during 5 days per week over a period of 6 weeks. A reminder of 8 training sessions was carried out before the post-treatment CB test, whereas the HAR was maintained. Statistical analysis was done with SAS, using a GENMOD procedure for training sessions and TTEST to compare the effect of the HAR on the results of the CB tests between GH and BH. During the remainder of 8 training sessions, pigs were able to discriminate between R and U bucket from session 4 onwards ($P=0.013$) and no difference in learning was found between GH and BH ($P=0.419$). To compare the effect of the HAR on the results of the CB tests between GH and BH, the mean of the difference of the time to contact the bucket between the two CB tests was calculated, GH (21.5 ± 14.30) and BH (12.0 ± 4.26), and no significant result was found ($P=0.558$). The lack of effect of HAR on CB could be due to the fact that the bad management used on BH room was not sufficiently distressing, to the low sensibility of the technique to assess differences in emotional state or to the large variability among individual pigs.

Acquisition of flavour preferences in pigs by interactions with pigs than previously drunk flavoured protein solutions

Jaime Figueroa[1,2], David Solà-Oriol[2], José Francisco Pérez[2] and Xavier Manteca[2]
[1]Universidad de Chile, Animal Production, Av. Santa Rosa 11735, La Pintana, Santiago, Chile, 8820000, Chile, [2]Universitat Autònoma de Barcelona, Departament de Ciència Animal i dels Aliments, Campus UAB, 08193 Bellaterra, Spain; avier.manteca@uab.cat

Social transmission of flavour preferences has been reported in several species when a demonstrator animal that has previously eaten a flavoured feed had the opportunity to interact with a naïve conspecific (observer). The aim of this experiment was to study if observer pigs could learn a flavour preference after interacting with a demonstrator that previously experienced flavours in protein solutions instead of feed. A total of 96 nursery pigs were used. Animals were allocated after weaning (28 d old) in 12 pens (8 pigs/pen). Four animals per pen were randomly selected three weeks after weaning (49 d old) to act as observers and 4 as demonstrators. Demonstrator animals were temporary moved to an empty pen where a protein solution (Porcine Digestive Peptides; 4% PDP) was offered with the inclusion of 0.075% of aniseed (PA) or garlic (PG) flavours for 30 min. After that, demonstrator piglets were mixed again with the observer animals for another 30 min. Flavours were counterbalanced across pens to act as the social (given to demonstrators) or control flavour. A choice test between aniseed and garlic flavoured solutions was performed in each observer group after the interaction time. Solution intakes were measured after 30 minutes. Data was analysed by using the GLM procedure of SAS®. No differences were observed between demonstrator's intake of PA or PG. After social interactions, observer pigs showed a higher intake of flavours previously experienced by demonstrator conspecifics over control flavours (648 vs. 468 ml; SEM 61.36, P=0.05). As suggested in previous experiments, our results show that pigs like other mammals are able to prefer flavours previously learned by social interaction. Protein solutions as well as feed could be an appropriate way to create new preferences by social learning even with its higher transit speed through oral cavity before social interaction occurs.

An exploration-based test for assessing affective state in mice

Janja Novak, Luca Melotti, Jeremy D. Bailoo and Hanno Würbel
Division of Animal Welfare, VPH Institute, Vetsuisse Faculty, University of Bern, Länggassstrasse
120, 3012 Bern, Switzerland; janja.novak@vetsuisse.unibe.ch

In recent years, a number of studies have applied tests of cognitive bias in animals as a means of inferring emotional states. These tests usually require extensive training which in turn can confound results. We modified a test of exploratory behaviour for mice, which is quicker to implement and is a promising measure of affective state. CD-1 mice (n=24) were trained in an eight arm radial maze, where reaching ends of two arms (positive arms) turned the overhead light off and released a food reward, while reaching ends of two opposite arms (negative arms) turned the overhead light back on and triggered a burst of white noise. The four remaining arms (ambiguous arms) were located between positive and negative arms and were closed during training. Over six training sessions, the proportion of time spent in the positive arms increased ($F_{1,23}$=4.55, P<0.05), indicating an overall preference for the positive arms across days of training. Overall, mice spent 66.57±15.8% of time in positive arms. Based on the preference for the positive arms, we hypothesised that time spent in ambiguous arms would be influenced by associations made with the positive and negative arms. In the testing session, mice were left to explore the maze with all eight arms open. Consistent with training data, mice showed a clear preference for positive arms during testing (t_{24}=3.29, P<0.05). They also spent more time in the ambiguous arms adjacent to positive arms compared to the ambiguous arms adjacent to negative arms (t_{24}=3.05, P<0.05). These differences suggest a clear discrimination between the different ambiguous arms. While the task still needs further validation and replication, our results suggest that this test could be a relatively quick and practical means of assessing cognitive biases in mice.

An enriched arena for lambs – anticipatory behaviour for access and behavioural responses when access was denied;

Aikaterini Zachopoulou, Claes Anderson and Lena Lidfors
Swedish University of Agricultural Sciences, Department of Animal Environment and Health, P.O. Box 234, 532 23 Skara, Sweden; aiza0001@gmail.com

This study aimed to investigate whether lambs show anticipatory behaviours in a holding pen prior to entering a known arena and how their behaviour is affected if access to arena is denied. Twenty pair-housed male lambs (10-13 weeks old) were exposed in chronological order to the following sessions. HP-sessions (three repetitions): lambs were led in pairs into a holding pen (2.7 m^2) and remained there for 5 minutes before returning to their home pen (6 m^2). A-sessions (three repetitions): Same as during HP-sessions, but following the holding pen, lambs entered into an arena (22 m^2) containing two hanging chains, one ball and one platform for 15 min before returning to their home pen. H-sessions (once): Same procedure as the HP-sessions, but under the assumption that lambs anticipated arena access. Prior to HP, lambs were habituated to entering the holding pen (five repetitions). Prior to A and H, lambs were conditioned to anticipate entering the arena (five repetitions of the A procedure). Behaviours were video recorded in holding pen, arena and home pens (ten minutes following their return) and extracted (Observer XT 11.5, Noldus Technology, Wageningen, the Netherlands). Data were analysed using Wilcoxon Signed ranks test. Between HP and A in the holding pen, no differences were found between behaviours indicative of anticipation (walking, exploring (sniffing pen), behavioural transitions (number of shifts between behaviours), play (locomotor and social play) – all n.s.). Compared to HP in the home pens, H resulted in more (% of duration) foraging (HP: 47%±6.06; H: 66%±4.88; P<0.01) and less ruminating (HP: 21%±4.96; H: 4%±2.31; P<0.01) and sitting (HP: 21%±5.15; H: 12%±4.44; P<0.05). In conclusion, male lambs did not appear to express behaviours indicating anticipation before entering the arena (HP vs. A), but when access to the arena was denied (H) different reaction was observed compared to the HP sessions.

Will an anticipation period facilitate play performance in lambs?

Claes Anderson, Aikaterini Zachopoulou and Lena Lidfors
Swedish University of Agricultural Sciences, Department of Animal Environment and Health,
P.O. Box 234, 532 23 Skara, Sweden; claes.anderson@slu.se

To our knowledge, no studies have looked into whether anticipation prepares an animal for opportunities to play. We predicted that lambs given an anticipation period prior to opportunities to play would result in higher levels of play compared to when they were not offered an anticipation period. One focal lamb from each of ten pairs of male lambs was studied. The lambs were first habituated to enter into a holding pen (eight repetitions), a box adjacent to a larger arena. Following this, the lambs were then repeatedly (five repetitions of training) exposed to entering into the holding pen for five minutes, to induce anticipation, followed by 14 minutes in a larger arena containing toy objects. After training, lambs underwent the same procedure (T1) and their behaviours were recorded (three repetitions) continuously in the holding pen and arena. Lambs were then led straight into the arena for 14 minutes without staying in the holding pen (T2) and behaviours in the arena were recorded (three repetitions). In T2, lambs spent a higher proportion of the time playing (33.8%±4.1) compared to T1 (27.1%±2.4, S=18.5, P<0.1). During the first seven minutes (half of the time in the arena), T2 spent significantly more time (37.2%±5.1) playing compared to T1 (30.1%±3.2; S=19.5, P<0.05). During the last seven minutes, there was no difference (T1=24.5%±2.0, T2=30.5%±3.6). The heart rate variability of the focal lambs was also recorded but has not been analysed yet. Contrary to our prediction, an anticipation period did not result in more play behaviour.

Goats alter their behaviour depending on the head orientation of a human but don´t use head direction as communicative cue

Christian Nawroth[1], Eberhard Von Borell[1] and Jan Langbein[2]
[1]Institute of Agricultural and Nutritional Sciences, Martin-Luther-University Halle-Wittenberg, Department of Animal Husbandry & Ecology, Theodor-Lieser-Str. 11, 06120 Halle (Saale), Germany, [2]Leibniz Institute for Farm Animal Biology, Institute of Behavioural Physiology, Wilhelm-Stahl-Allee 2, 18196 Dummerstorf, Germany; christian.nawroth@landw.uni-halle.de

In a first experiment, we investigated the effect of different body postures of a human on the behaviour of individually tested dwarf goats (n=11; 2-3 years of age) in a food-requesting paradigm. The experimenter was sitting opposite to the subject and both were separated by a mesh. He placed a food reward inaccessible for the subject on a sliding board behind the mesh and waited 30 sec until sliding it towards the mesh to deliver the reward. During this time, he engaged in one of the four condition: 'Control' (head and body oriented towards subject), 'Head' (only head oriented away from subject), 'Back' (head and body oriented away from subject), and 'Out' (experimenter left the room for 30 sec). We recorded the time a subject was oriented towards the mesh and the experimental set-up and engaged in (1) active anticipatory behaviour (nervous tripping in front and repeatedly snouting through the mesh) and (2) standing alert (standing motionless in front of the mesh while facing the experimental set-up). SAS 9.3 was used for statistical analysis. We found that the level of subjects´ active anticipatory behaviour was positively correlated with the level of attention the experimenter was paying to the subject (linear mixed model; n=9; post-hoc comparisons P<0.05), while standing alert increased when the experimenter was present, but showed reduced attention to the subject ('Back' and 'Head' condition) compared to the 'Control' and 'Out' conditions (linear mixed model; n=9; post-hoc comparisons P<0.05). In a second experiment, an experimenter provided different human-given cues (Touch, Point, Gaze, and Control) that indicated the location of a hidden food reward which was placed beneath one of two containers. We found a significant interaction of test condition and time of test (sessions 1-3 vs. sessions 4-6) on performance (generalized linear mixed model; P=0.011). Post-hoc tests revealed that subjects performed better in the 'Touch' and 'Point' condition compared to the 'Control' condition in sessions 1-3 (all P<0.01), and better in the 'Touch' and 'Point' condition compared to 'Gaze' and 'Control' in sessions 4-6 (all P<0.01). We conclude that goats are able to differentiate between different body postures of a human, including head orientation, but – despite their successful use of various physical human cues – fail to use human head direction spontaneously as information in a food-related context.

Moving on to a new coexistence – understanding equine socio-cognition in the horse-human relationship

Francesco De Giorgio and José Schoorl

Learning Animals, Institute for Zooanthropology, Hool 62, 5469 KC, the Netherlands; info@learning-animals.org

With a growing interest and awareness for animal quality of life, and understanding of the animal-human relationship, it becomes more and more necessary to develop a practical understanding of the knowledge related to topics with an important impact on animal quality of life, as for example affiliative behavior and cognitive abilities. This is especially the case when it comes to animals living closely in relationship with human, and where a long history of traditions is a potential hindrance for application of new insights in both day-to-day interaction as well as in research context, as for example regarding horses in the horse-human relationship. The principal aim of this paper is to explain the possible interference of human belief-systems within applied ethology regarding horses. Equine cognition has been shaped by the evolutionary process, both by the environmental challenges and horses' complex social dynamics, resulting in strong socio-cognitive characteristics. The understanding of these characteristics should be the foundation of any interaction with human. Starting with accepting their social needs for an affiliative environment whenever undertaking unknown interactions with human. This means first of all understanding the impact of training techniques, as they might reduce the cognitive-relational abilities and disturb the behavioral expressions in the relationship with humans, as they are based on hierarchical way of interactions. For their quality of life and an ethical interaction with human, horses should be able to actively participate in a socio-cognitive environment having the human as a partner in shared experiences. It means considering in the day-to-day practice, the horse's socio-cognitive abilities to elaborate cognitive maps, to search for information, to elaborate and process knowledge, to follow his own inner motivation, to express emotions or intentions, to solve problems, to adapt to changes, and especially to develop relationships based on affiliative expressions. All these elements form the core of any experience, leaving behind the assumption that a horse should be trained and conditioned. These baseline indications are reassumed in the cognitive-zooanthropologic model, where the affiliative and cognitive abilities of animals are central as they should be taken into account for the development of a reciprocal horse-human relationship, that enables positive shared experiences for both.

Pair associations of Asiatic black bears (*Ursus thibetanus*) rescued from bear bile farms

Jessica Frances Lampe[1], Fernanda Machado Tahamtani[2], Nicola Field[3] and Heather Bacon[4]
[1]University of Bern, Vetsuisse, Animal Welfare Division, Länggassstrasse 120, 3012, Switzerland,
[2]Norwegian School of Veterinary Science, Department of Production of Animal Clinical Science,
Animal Welfare Group, 72 Postboks 8146 Dep., 0033 Oslo, Norway, [3]China Bear Rescue Centre,
Animals Asia Foundation, China Bear Rescue Centre, Longqiao, Xin Du District, Chengdu,
Sichuan 610505, China, P.R., [4]Jeanne Marching International Centre for Animal Welfare
Education, University of Edinburgh, The Royal (Dick) School of Veterinary Studies, Easter Bush
Veterinary Centre, EH25 9RG Roslin, United Kingdom; jessica.lampe@vetsuisse.unibe.ch

The study reports evidence that captive Asiatic black bears (*Ursus thibetanus*), usually considered solitary, can develop and maintain lasting pair relationships by interacting in either affiliative or agonistic bearings, within and across gender. Eight pairs of bears (Animals Asia Sanctuary, Longqiao, China) were observed, living in 3 different group enclosures housing 9 to 16 bears. Previous observations by staff had identified these pairs as showing lasting, i.e. yearlong, mutual associations. Although other group members also interacted in affiliative or agonistic ways and sometimes formed short-term relationships, they did not show long-term associations. An ethogram was developed to document 10 different types of social behaviours, split into 5 behaviours with positive (affiliative) and 5 with negative (agonistic) valence. Positive: play wrestling without vocalizations, nuzzling/pawing, tongue clucking, resting in close proximity, protective behaviour; negative: fighting with vocalizations, gaping without vocalizations, growling, chasing away, avoiding/escaping. These types of behaviours were shown to differ significantly in their levels of importance (based on durations: $\chi^2_4=8.9$, P=0.04 [affiliative]; $\chi^2_4=9.4$, P=0.04 [agonistic]). Pairs interacted exclusively either in an affiliative (n=4) or agonistic (n=4) way. The positively interacting pairs spent significantly more time on their affiliative behaviours than the negatively interacting ones on their agonistic behaviours ($F_{1,6}=8.24$, P=0.02). Morning and afternoon durations of social behaviour did not differ significantly (Z=-0.140, P=0.94). Not having been tapped for bile had a significant positive (increasing) effect on the overall duration of affiliative behaviours (Kruskal-Wallis, $\chi^2_1=4.26$, P=0.04, based on comparison of pairs in which none, one or both partners had been tapped for bile). Although Asiatic black bears are considered a solitary species, the study showed that within larger social groups these bears are able to maintain pair associations with either positive or negative behavioural repertoires.

Tear staining as an indicator of stress caused by dominance rank and housing in pigs

Moriah Hurt, Judith Stella and Candace Croney
Purdue University Center for Animal Welfare Science, Animal Sciences and Comparative Pathobiology, 915 W. State St., G-405 Lilly Hall, West Lafayette IN 47907, USA; jstella@purdue.edu

Tear staining in pigs has been suggested to be a useful non-invasive indicator of distress in pigs and may provide information about the relative activation of the HPA and SAM axes. It has been shown to be negatively correlated with plasma cortisol concentration and positively correlated with a sympathetic nervous system indicator (low frequency power/high frequency power). This study aimed (1) to investigate differences in tear staining between dominant and submissive pigs housed in pairs and (2) to evaluate tear staining as an indicator of distress in pigs kept in pair and single housing conditions. Ten, eight week old Landrace × Duroc barrows were randomly allocated to pairs and housed together in indoor pens 2.37 m^2. Each pig was designated as dominant or submissive based on the outcome of a dominance test conducted with each pair. Once weekly for 8 weeks, an experimenter scored level of tear staining on each eye for each pair of pigs using a scale of 0-5 with 0 indicating no staining and 5 indicating excessive staining. After 5 weeks of pair housing, the growth of the pigs and increased fighting bouts led to a decision to singly house pen mates adjacent to each other. Scores were analyzed by 2-WAY ANOVA to test the effect of (1) dominance and week and (2) housing and week. Dominance rank was not significant (P=0.8). However, the effect of week was statistically significant (P<0.0001). Analysis of the effect of housing condition and week showed neither housing (P=0.4) nor week (P=0.1) was statistically significant but the interaction between week and housing was statistically significant (P<0.0001). A statistically significant interaction between week and housing was found for tear-staining in both the left (P=0.002) and the right eyes (P=0.0008). Rank did not impact tear staining. However, housing condition did appear to affect staining with reduced staining occurring when pigs were singly housed. It is possible that the reduced floor space and increased aggression between pigs that occurred over time in pair housing caused distress that manifested in increased staining, which decreased when those stressors were mitigated. Tear staining may indeed be a reliable indicator of social stress in pigs as a function of housing.

The social network of the Japanese macaques (*Macaca fuscata*) at the Highland Wildlife Park

Lauren M Robinson[1] and Rob Thomas[1,2]
[1]University of Edinburgh, Old College, EH8 9YL, United Kingdom, [2]Royal Zoological Society of Scotland, 134 Corstorphine Rd, EH12 6TS, United Kingdom; l.robinson@ed.ac.uk

Zoo animals are regularly transferred to different zoos where they are added to current exhibits. It is rare that different groups of animals that are brought together in a new exhibit. We were able to study such a group when we researched the social network of the Japanese macaques (*Macaca fuscata*) at the Highland Wildlife Park in Kincraig, Scotland. These macaques were brought together in 2007 from three European zoos. At the time of data collection the 18 macaques had been together for five years. The macaques live in a 1.35 ha enclosure giving the animals enough room to choose their preferred social partners and avoid unwanted social interactions. Therefore, we chose to use social network analysis to the study the macaque social relationships looking for differences in social relationships based on troop-of-origin. We gathered data on grooming behaviour and conspecific proximity tendencies. Proximity data was gathered using scan sampling; grooming data was collected by focal sampling different exhibit areas. We found a significant correlation between proximity and grooming (r_{s18}=0.376, P<0.01). We were able to collect significantly more proximity observations than grooming therefore we used the proximity data to map the social network. A dendogram of the proximity data was created to visually assess the social sub groups within the troop. It was found that after five years together the macaques heavily favoured the company of their original troop members. The three strongest connections within the troop were between the mothers and their offspring who were aged one, two, and eight years. The youngest macaque displayed the most flexibility in her social relationships and had the most connections to different conspecifics. Our results suggest that, while the macaques at the Highland Wildlife Park have lived together for five years, their social groups are still defined by their troop-of-origin.

Why does novelty increase locomotor play in some veal calves but decrease it in others?

Jeffrey Rushen and Anne Marie De Passille
University of British Columbia, Agassiz, BC V0M 1A0, Canada; jeffrushen@gmail.com

Locomotor play by calves is increased by novelty, which appears to contradict the suggestion that locomotor play is a sign of good welfare since novelty usually produces fearfulness in animals. We hypothesized that calves responding to novelty most with exploratory behaviour would show more locomotor play, while those responding to novelty with signs of fear would show less. In two experiments, we examined the locomotor play of 48 veal calves placed for 15 min in a novel arena (11×3.2 m) and measured the frequency of jumping (locomotor play) and the frequency of sniffing (as a measure of exploration) and vocalization, defecation, and the latency to enter the arena (as measures of fear). In experiment 1 (6 weeks of age), the jumping was correlated with sniffing ($r=0.46$; $P<0.05$) and negatively correlated with the latency to enter the arena ($r=-.64$; $P<0.01$). Defecation was not correlated with jumping. In experiment 2 (11 weeks of age), sniffing was positively correlated with jumping ($r=0.51$; $P<0.001$), while mooing vocalizations were negatively correlated ($r=-0.35$; $P<0.05$). In contrast, 'baaock' vocalizations were positively correlated with jumping ($r=0.35$; $P<0.01$). Jumping increased with age (Wilcoxon test; $P<0.001$), as did the frequency of sniffing (Wilcoxon test; $P<0.001$), and the size of the change with age in jumping was correlated with the size of the change in sniffing ($r=0.54$; $P=0.01$). The effects of novelty and age on locomotor play appear to be mediated by the calves' levels of exploratory behaviour, with calves that show more exploration doing more play. Among younger calves, fear reduces playfulness, but the increase in locomotor play with age does not appear to be due to a reduction in fearfulness.

To what extent do men and women believe animals can experience emotions?

Nancy Clarke[1], Liz Paul[2] and David Main[2]
[1]*World Society for the Protection of Animals, Education, 222 Gray's Inn Road, London, WC1X 8HB, United Kingdom,* [2]*University of Bristol, Animal Welfare and Behaviour Group, School of Veterinary Sciences, Langford House, Langford, Bristol BS40 5DU, United Kingdom; nancyclarke@wspa-international.org*

Research demonstrates small but consistent gender differences in attitudes towards animals, with females reporting more positive feelings and concern for animals than males. Significant gender differences have also been demonstrated in the personality traits of 'empathizing' and 'systemizing'. The present studies were conducted to investigate whether these traits can account for reported gender differences in attitudes to animals, and/or for within-gender variation in such attitudes. In Study 1 (n=211), a convenience sample of undergraduates completed an online questionnaire, rating a range of animal species' capacity for six emotions on a 9-point scale, the Empathy Quotient (EQ-Short) and Systemizing Quotient (SQ-Short). In Study 2, a group of animal protectionists (n=223) and a non-animal protectionist comparison group of office workers (n=83) completed the same online questionnaire. In Study 1, relative to males, females had significantly higher empathizing and lower systemizing tendencies and rated animals more highly as having capacity for emotions. In Study 2, Animal Protectionists demonstrated significantly higher belief in animal capacity for emotions, higher EQ scores and lower SQ scores relative to the Office Worker comparison group. Both female and male Animal Protectionists demonstrated significantly higher belief in animal capacity for emotions relative to their respective counterparts in the Office Worker comparison group. Male animal protectionists had significantly higher EQ scores than the males within the Office Worker comparison group and there were no significant differences in the male's SQ scores between these groups. Overall, the results suggest that individual differences in empathizing and systemizing tendencies account minimally for gender differences in belief in animal capacity for emotion, but that empathizing tendencies might mediate males, but not females, belief in animal capacity for emotions. Other factors relating to how people self-report empathy but also their belief in animal emotions require further investigation.

How do beef producers perceive pain in cattle?

Ingela Wikman[1], Ann-Helena Hokkanen[1], Matti Pastell[1,2], Tiina Kauppinen[1], Anna Valros[1] and Laura Hänninen[1]
[1]Research Centre for Animal Welfare, Faculty of Veterinary Medicine, P.O. Box 57, 00014 University of Helsinki, Finland, [2]Department of Agricultural Sciences, P.O. Box 28, 00014 University of Helsinki, Finland; ingela.wikman@helsinki.fi

Attitudes play a key role when assessing and treating pain in animals. Dairy procures´ and veterinarians´ perceptions of cattle pain have been found to be affected by persons´ age, work experience and gender but little is known about beef producers. Thus we aimed at studying beef producers´ attitudes of the painfulness of different cattle diseases and disbudding practices. A questionnaire about cattle pain and disbudding practices was sent to 1000 Finnish beef producers with a response rate of 44% (n=439). Producers graded their attitudes about disbudding on a five-point Likert scale and perception about cattle pain on an eleven-point scale. A principal component analysis (PCA) was used to assess the factor loadings on four previously found factors describing stakeholders perception of cattle pain: Factor I ('taking disbudding pain seriously'), Factor II ('sensitivity to pain caused by cattle diseases'), Factor III ('ready to medicate calves myself') and Factor IV ('pro horns'). Factor loadings were tested for differences between genders (men, n=293, women, n=135) with Mann Whitney U-tests, and Kruskal-Wallis tests were used to assess differences between producers´ age (≥55 years, n–99, 40-54 years, n=214 and ≤39 years, n=111) and work experience (maximum of 5 years, n=152; 6-10 years, n=108; 11-20 years, n=88; and >20 years; n=77). Possible pairwise comparisons were tested with Mann-Whitney U-tests using Bonferroni corrections. Beef producers´ perceptions loaded to all previously found factors. Factor loadings differed between genders, producer´s work experience and age (P<0.05 for all): Female producers had higher loadings than males for Factors I, II and III. Producers over 55 years had higher loadings for Factors II and IV than producers 39 years of age or younger. Producers had been working over 20 years had higher loadings for Factor IV than the ones been working for maximum of 5 years or 11-20 years. We found that older and experienced beef producers show more positive attitudes to cattle with horns than younger or less experienced ones. Female beef producers assess cattle pain higher and have more positive attitudes to medicating pain in cattle. Older beef producers were sensitive to cattle pain probably reflecting their experience of cattle diseases which have been previously found to be connected with a higher empathy. As we have shown previously in dairy producers; gender, age and work experience are affecting also beef producers´ attitudes to cattle pain. However the cultural effects to factor loadings cannot be ruled out and further studies are needed.

Do disbudded calves benefit from five-day pain medication?

Juha Hietaoja, Suvi Taponen, Matti Pastell, Marianna Norring, Tiina Kauppinen, Anna Valros and Laura Hänninen
Research centre for animal welfare, P.O. Box 57, 00014 University of Helsinki, Finland;
laura.hanninen@helsinki.fi

Disbudding causes pain, but the needed length of post-operative pain medication or the possible differences between disbudding with heat cauterization vs caustic paste are not known. We studied the differences between these methods and the effect of prolonged treatment with a non-steroidal anti-inflammatory drug (NSAID) on dairy calves' activity and milk-feeder visits. We performed a 2×2 factorial design study on two-week old dairy calves weighing (mean ± SE) 51±1 kg. We disbudded calves during sedation with either heat cauterization (n=24) or caustic paste (n=23). Cauterized calves were also given local anaesthesia (lidocain). All calves got NSAID (ketoprofen) orally (4 mg/kg) before sedation (detomidine), and the NSAID5 during four more days (n=24). NSAID1 got equivalent volume of water (n=23). All calves were sham-disbudded 2 days prior (day -2) disbudding (day 0). We registered calves activity and milk-feeder visits with accelerometer and automatic-milk-feeder from day -2 to day 8. Calves were weighed days -2, 0, and 10, and calculated number of nutritive and non-nutritive visits, consumed milk, and number and duration of resting bouts and total durations for daily resting and activity. The differences between treatments were analysed with repeated sampling linear mixed models. Disbudding method did not affect resting behaviours but NSAID tended to decrease daily overall activity (P<0.06): NSAID5 spent less time active than NSAID1 (473±24 vs. 530±22). Treatment × day interactions were found for the number of all and non-nutritive feeder visits (P≤0.05 for all): NSAID5 had more visits than NSAID1 on days 3 and 4 (19±1 and 14±1 vs. 16±1 and 12±1, respectively) and non-nutritive visits on day 3 (10±1 vs. 5±1). Cauterized calves drank quicker than caustic paste disbudded ones (0.7±0.03 l/min vs. 0.6±0.03 l/min; P=0.015). Cauterized calves grew overall better than paste disbudded calves (0.93±0.05 kg/d vs. 0.88±0.05 kg/d) and NSAID5 grew better than NSAID1 (1.0±0.05 kg/d vs. 0.6±0.05 kg/d; P<0.05 for both). Also a medication × method interaction was found (P<0.05): cauterized NSAID5 calves grew better than cauterized NSAID1 calves (1.1±0.1 kg/d vs. 0.8±0.1 kg/d), no difference was found within caustic paste disbudded animals (0.9±0.1 kg/d vs. 0.9±0.1 kg/d, respectively). Amount of consumed milk did not differ between treatments. Disbudded calves might benefit from a five-day NSAID medication, as it increased growth, reduced overall activity and increased number of feeder visits. Caustic paste calves had decreased growth and drinking speed, with no effect of NSAID.

Out of sight, out of mind? Effects of structural elements in the lying area on behaviour and injuries of horned dairy cows

Susanne Waiblinger[1], Maria Peer[1], Claudia Schneider[2], Anet Spengler[2] and Christoph Menke[3]
[1]University of Veterinary Medicine Vienna, Department of Farm Animals and Veterinary Public Health, Institute of Animal Husbandry and Animal Welfare, Veterinärplatz 1, 1210 Vienna, Austria, [2]Research Institute of Organic Agriculture, Animal husbandry and consulting departments, Ackerstrasse 21, 5070 Frick, Switzerland, [3]Association for Animal Welfare Research, VEAT, Aureliaweg 16, 93055 regensburg, Germany; susanne.waiblinger@vetmeduni.ac.at

Deep-litter and straw-flow systems are advantageous over cubicle loose housing with regard to various aspects of dairy cow welfare, but disturbances of lying animals can be higher, especially if animals have horns. Structural elements might reduce social disturbances and enhance resting and consequently cleanliness. We investigated this question on five commercial farms keeping horned dairy cows in a deep-litter system. Data were collected in situations without and with a structural element present (three wooden walls, each 2.50 m long, 1.50 m high, arranged in a Y-shape) for four days each. Social behaviour was recorded directly and continuously for 4 h per day, resting behaviour by video for 24 hours per day with scan sampling every 5 min. Cleanliness and injuries were assessed at the beginning and end and the difference was calculated. The proportion of lying for 24 h and at night was higher with the structural elements present (at night: 48%) than without (44,5%; linear mixed model: P<0.001). This was particularly true for high-ranking and middle-ranking animals (interaction rank × structure: P=0.007), for which lying time increased by about 1 h on average. The strength of the effect, however, varied considerably between individual farms. Agonistic interactions in total occurred less often in the resting area with the structural element present (mean: 7.81 interactions/h) than without structure (8.23/h, P<0.05), mainly due to difference in interactions without body contact, while with body contact no difference was found. However without the structural element present, the increase in integument lesions (total) of the cows was less than with it (P<0.05). Animals were less dirty with the structural element present than without (P=0.008). The results indicate a positive effect of the presence of a Y-shaped structural element in the resting area on lying time, social behaviour and cleanliness of horned dairy cows, but a negative one on injuries. In sum the use of structural elements in straw yard or comparable systems has the potential to improve the welfare of horned dairy cows.

Liquid nitrogen cryosurgery as an alternative to hot-iron disbudding of dairy calves

Mairi Stewart, Suzanne Dowling, Karin Schütz, Vanessa Cave and Mhairi Sutherland
AgResearch Ltd, Private Bag 3123, Hamilton 3124, New Zealand; mairi.stewart@interag.co.nz

Alternatives for painful husbandry procedures are required to demonstrate high standards of welfare on farm. This study investigated the efficacy of cryosurgery (application of liquid nitrogen to induce localised cell destruction) to prevent horn bud growth and its potential as an alternative to hot-iron disbudding of dairy calves. To assess efficacy of the procedure, 274 dairy calves, 2-5 days old, on two commercial farms (167 calves on one farm and 107 on the other), received cryosurgery; liquid nitrogen applied to the primordial horn bud area (30 mm diameter), using a commercial liquid nitrogen applicator (Brymill Cry-Ac B-700) for 10 s on one side and 15 s on the other, with side × timing randomised. REML was used for statistical analysis. The 15 s treatment had a higher success rate for preventing horn bud growth than 10 s (47 vs 30±3%, 15 vs 10 s respectively, P<0.001). A separate pilot study used thirty four 2-4 day old dairy calves to investigate lying behaviour (recorded continuously using Onset Pendant G data loggers) during the day of treatment and for 3 days post-treatment. Calves were allocated to either (1) Handling control (n=11): simulated handling only, (2) Cryosurgery (n=12): Liquid nitrogen applied (10 s) over horn bud area (20 mm diameter) or (3) Disbudded (n=11): horn buds removed with cautery iron. Mean lying (min/hour) and SEM were calculated. REML was used for statistical analysis. On the day of treatment, lying was less (P<0.01) for calves Disbudded (48.2±1.1 min/h) compared to Cryosurgery (52.3±1.1 min/h) or Controls (50.7±1.1 min/h). There were no differences in lying times between Cryosurgery and Controls. In summary, due to the minimal tissue damage, cryosurgery did not appear to cause the extensive inflammatory responses that disbudding produces. The higher lying times in the cryosurgery calves suggest they may have experienced less discomfort. Other responses would be worth investigating and further studies will be undertaken to assess the welfare implications of cryosurgery compared to disbudding. The duration of the application needs to be longer to improve the success rate of horn bud cell destruction. This has also been confirmed with recent histological examinations. This technique has potential as a more humane alternative to disbudding dairy calves but further refinement of the technique is required.

Features of the equine pain face

Karina Bech Gleerup[1], Pia Haubro Andersen[2], Casper Lindegaard[3] and Björn Forkman[1]
[1]University of Copenhagen, Department of large animal sciences, Gronnegaardsvej 8, 1870 Frederiksberg, Denmark, [2]Swedish University of Agricultural Sciences, Department of Clinical Sciences, Klinikcentrum, Ulls väg 12, 750 07 Uppsala, Sweden, [3]Helsingborg Regional Animal Hospital, Equine department, Bergavägen, 250 23 Helsingborg, Sweden; kbg@sund.ku.dk

In many animals moderate pain induces only minor changes in their behaviour. Recent research has shown that facial expressions may be a sensitive indictor of pain in some species. The aim of this study was to investigate the possible existence of an equine pain face. Clinically relevant pain was induced by two methods, a tourniquet-induced ischemia and a capsaicin-induced inflammatory pain in six horses in a semi-randomized controlled cross-over trial. Both methods induced moderate, short-lasting and fully reversible pain. Pain induction was used to ensure comparable levels of pain. All horses received both pain inductions twice, once with and once without the observer present. During all sessions the horses were video filmed and the close-up video recordings of the faces were analysed on a separate occasion and pain state was quantified using a composite measure pain scale. Both methods produced altered behaviour interpreted as pain, in all horses. No physiological parameters changed significantly, further indicating that the level of pain was only moderate. Alterations in facial expressions were observed in all horses and facial expressions representative for baseline and pain expressions were condensed into explanatory illustrations. An equine pain face comprising 'low' and/or 'asymmetrical' ears, an angled appearance of the eyes, medio-laterally dilated nostrils and tension of the lips, chin and certain mimetic muscles can be recognized in horses during induced pain. The presence of the observer did not influence the facial expression during pain sessions but the horses increased certain contact-seeking behaviors. These results suggest the equine pain face as a useful tool for pain evaluation in horses with moderate pain.

Farmers' and veterinarians' perceptions of lameness and pain in sheep

Carol S Thompson[1], kenneth M D Rutherford[1], joanne Williams[2] and adroaldo J Zanella[3]
[1]Scotland's Rural College (SRUC), Animal Behaviour & Welfare, Kings Buildings, West Mains Road, Edinburgh, EH9 3JG, United Kingdom, [2]The University of Edinburgh, School of Health in Social Science, Teviot Place, Edinburgh, EH8 9AG, United Kingdom, [3]Universidade de São Paulo, Departamento de Medicina Veterinária Preventiva e Saúde Animal, Faculdade de Medicina Veterinária e Zootecnia, Av Duque de Caxias Norte, 225, 13635-900, Pirassununga, SP, Brazil; carol.thompson@sruc.ac.uk

Accurate identification of disease and pain is paramount to good animal welfare. This study aimed to assess how farmers and veterinarians perceived lameness and pain in sheep. Movie clips (20 seconds each) of four sheep with varying levels of lameness (one 'sound', one 'mildly lame' and two 'moderate/severely lame') were shown to farmers (n=68) and veterinarians (n=46) in order to investigate how they perceived lameness and its associated pain. After each clip, participants completed a short questionnaire, which asked them to rate, using a 100 mm visual analogue scale, the level of: (1) lameness (L), (2) pain (P) they felt the sheep was experiencing, and (3) their own emotional (E) response. Data were analysed (Genstat 15) using REML and Spearman Rank Correlations. Strong positive correlations were found between ratings of lameness, pain and emotional reaction for both farmers and vets (Farmers: L vs P: r=0.93, P<0.001; L vs E: r=0.81, P<0.001; P vs E: r=0.84, P<0.001. Vets: L vs P: r=0.95, P<0.001; L vs E: r=0.88, P<0.001; P vs E: r=0.91, P<0.001). As intended, there were significant differences between sheep in all variables (L: W=1529.7, P<0.001; P: W=1370.24, P<0.001; E: W=846.54, P<0.001). Analysis revealed no significant effect of profession on L (W=0.39, P=0.53), P (W−0.65, P=0.42) or E scores (W=1.86, P=0.18), and no interactions were found between 'Sheep' and 'Profession'. However subsequent post hoc analysis revealed that although farmers and vets scored the 'sound' and 'mildly lame' sheep similarly, vets scored the two 'moderate/severely lame' sheep significantly higher than farmers for all three variables (e.g. for the most lame sheep (mean ± SE): L: Farmers=81.5±1.7, Vets=85.9±1.8, W=6.14, P=0.015; P: Farmers=74.3±2.3, Vets=80.9±1.9, W=7.57, P=0.007; E: Farmers=60.3±3.1, Vets=71.8±3.1, W=8.06, P=0.005). These results indicate that participants view lameness as a painful condition. However, for the moderate/severely lame sheep, the ratings provided by vets were significantly higher than those provided by farmers. This may have important implications for decision making surrounding treatment. Farmers' threshold for treatment may be higher than vets, which could have consequences for the health and welfare of lame sheep.

Bioavailability and efficacy of orally administered flunixin, carprofen and ketoprofen in a pain model in sheep

Danila Marini[1,2], Joe Pippia[3], Ian Coldtiz[2], Geoff Hinch[1], Carol Petherick[4] and Caroline Lee[2]
[1]The University of New England, Armidale, 2350, NSW, Australia, [2]CSIRO, Armidale, 2350, NSW, Australia, [3]PIA PHARMA Pty Ltd, Gladesville, 2111, NSW, Australia, [4]The University of Queensland, St Lucia, 4067, QLD, Australia; danila.marini@cirso.au

The pain from routine husbandry practices performed on sheep can last several days and sheep often don't receive therapeutic interventions to provide pain relief. Attractive candidates for long-acting pain relief are non-steroidal anti-inflammatory drugs (NSAIDs). If NSAIDs can be shown to alleviate pain and inflammation when administered orally in sheep, they could be incorporated in feed, providing producers with a practical method to provide long-term pain relief in sheep. The aim of this research was to test the bioavailability and efficacy of carprofen, ketoprofen and flunixin administered orally using a lameness model (turpentine (0.1 ml) injected into one forelimb) developed to enable objective quantitative assessment of the analgesic, antipyretic and anti-inflammatory actions of NSAIDs in sheep. Sheep were randomised into four treatment groups; with a placebo group receiving saline and three groups given NSAIDs as an oral dose every 24 h for 6 days: carprofen (8.0 mg/kg), ketoprofen (8.0 mg/kg) and flunixin (4.0 mg/kg). Responses measured included force plate pressure, lameness score, skin temperature, limb circumference and haematology at 0, 3, 6, 9, 12, 24, 36, 48, 72, 96 h. Concentrations of NSAIDs in plasma were analysed by Ultra High Pressure Liquid Chromatography. Data were analysed using linear and nonlinear mixed effects models. The NSAIDs were bioavailable in sheep 2 h after oral administration with carprofen and flunixin reaching inferred therapeutic concentrations in blood by this time. Range measured was 30 and 80 µg/ml for carprofen, 2.6 and 4.1 µg/ml for flunixin and 2.6-4.1µg/ml for ketoprofen. Turpentine injection induced lameness, increased limb circumference, increased limb temperature and elevated neutrophil to lymphocyte ratio. Paradoxically, placebo sheep placed more weight on their forelimbs at 3 h ($P<0.05$, 7.56±1.26 kg) compared with other treatments (carprofen 3.19±1.05 kg, ketoprofen 1.08±0.48 kg, flunixin 4.61±0.97 kg). Animals receiving flunixin had lower lameness scores than placebo sheep at 12, 24 and 48 h ($P=0.017$, 0.029 and 0.025 respectively). There was no significant effect of NSAIDs on haematology variables. The study showed the NSAIDs were bioavailable following oral administration in sheep. Future research will further examine efficacy due to paradoxical responses in placebo sheep in this study.

Computer-based learning in animal pain for UK veterinary students: effect on learning and attitude towards animal pain

Karen Hiestand and Fritha Langford
The University of Edinburgh, Royal (Dick) School of Veterinary Studies, The University of Edinburgh, EH25 9RG, United Kingdom; karenhiestand@hotmail.com

Despite the importance of effective pain management for comprehensive care in both humans and other animals, inadequacies remain in pain recognition and treatment in medical and veterinary fields. Introducing and increasing education specifically about pain recognition, assessment and treatment into veterinary curricula has been recommended to redress these inadequacies. Computer-based learning (CBL) can circumvent constraints on teaching time, resource and expertise that continue to limit adequate pain education provision. CBL is increasingly used in veterinary education and the effectiveness of various online tools has been investigated with generally positive results. As part of the larger 'Animal Welfare Indicators (AWIN)' project, this study provided UK veterinary students with CBL learning interventions on the topic of feline chronic pain. A randomized, single-blinded controlled trial was used to compare effectiveness of CBL against traditional presentation of comparable material. Six UK veterinary schools participated with a response rate of 22% (n=649). Students were recruited and participated entirely online. Participants first completed a pre-questionnaire consisting of demographic information, knowledge about feline pain, confidence and experience evaluating feline pain and attitudes to animal pain (multiple species). Students were then randomly supplied with either a CBL tool or a pdf document. Post-questionnaires were completed to assess knowledge transfer, potential attitude change, attitude towards learning intervention style and perception of learning. As in previous work, CBL and traditional methods were found to result in comparable knowledge transfer. For example, respondents were 2.3 times more likely to correctly identify the most common cause of chronic pain after learning interventions (95% CI 2.06-2.56, P<0.001) with no statistical difference between intervention type. Retention of material was assessed after two months and respondents remained twice as likely to answer this question correctly (95% CI 1.83-2.338, P<0.001). Students responded more favorably to interaction with CBL compared to conventional didactic presentation of materials, illustrated by the perception that they learnt more ($\chi2=5.281$, DF=1, P=0.02) and that the way information was presented helped them to learn ($\chi2=17.24$, DF=1, P<0.001). Belief in various species ability to experience pain similarly to humans was shown to increase significantly after specific pain education, particularly in CBL respondents (Wald, F(1,629)=15.67, P=0.02). This study further validates use of CBL for veterinary students and confirms its appropriateness for specific pain education.

Potential behavioural indicators of pain in periparturient sows

Sarah H Ison and Kenneth M D Rutherford
Scotland's Rural College (SRUC), Animal Behaviour and Welfare, Roslin Intitute Building, Easter Bush, Roslin, Midlothian EH25 9RG, United Kingdom; sarah.ison@sruc.ac.uk

Spontaneous behaviour, often termed 'pain-specific', is frequently used in research assessing pain in pigs. No studies have recorded 'pain-specific' behaviour around farrowing. The aim of this study was to identify potential behavioural indicators of pain around farrowing. Twenty-five sows (Large White × Landrace) housed in crates were used in this study. Focal observations of 10 sows were made for five minutes every hour between 60 and 36 hours before farrowing. These 10, along with another 15 were observed from the birth of the first to last piglet and for five minutes every hour for 24 hours post-farrowing. Behaviour recorded included posture, piglet births and a set of other behaviours: (1) back leg forward duration (BLF: the sow pulls her back leg forward and/or in); (2) tremble duration (T: the body moves as if shivering); (3) back arch (BA: the sow stretches forming an arch with the spine); (4) paw (P: the sow scrapes a leg in a pawing motion); and (5) tail flick (TF: the tail rapidly moves up and down). Behaviour was analysed using GLMMs with two way comparisons of observation periods: pre, during, post 1, post 2, post 3 and post 4 as a fixed effect and sow in the random model. The behaviours T and P were not seen in pre-farrowing observations. BLF and BA were greater (P<0.001) during, and post-farrowing compared to pre-farrowing, and TF was more (P<0.001) frequent during compared to pre-farrowing. All behaviours except P were greater (P<0.003) during compared to post-farrowing, and BLF, T, and BA were greater (P<0.003) in the early post-farrowing period. These behaviours, which show individual variation, may be useful indicators of pain, which could be used to identify sows that may benefit from post-farrowing analgesia, potentially improving welfare and productivity.

Birds' emotionality modulates the impact of chronic stress on feeding behaviour

Angélique Favreau-Peigné[1], Ludovic Calandreau[2], Bernard Gaultier[3], Paul Constantin[2], Aline Bertin[2], Cécile Arnould[2], Alain Boissy[4], Frédéric Mercerand[3], Agathe Laurence[5], Sophie Lumineau[5], Cécilia Houdelier[5], Marie-Annick Richard-Yris[5] and Christine Leterrier[2]
[1]INRA, UMR791 Modélisation systémique appliquée aux ruminants, 75231 Paris, France, [2]INRA, UMR85 Physiologie de la Reproduction et des Comportements, 37380 Nouzilly, France, [3]INRA, UE PEAT, 37380 Nouzilly, France, [4]INRA, UMR1213 Herbivores, 63122 St Genès Champanelle, France, [5]Université Rennes, UMR6552, 35042 Rennes, France; angelique.favreau@agroparistech.fr

Chronic stress is a long-lasting negative emotional state which induces negative consequences on animals' behavior. This study aimed at assessing whether unpredictable and repeated negative stimuli (URNS) influence feeding behavior in quail, and whether this can be modulated by their emotionality. Two lines of quail divergently selected on their inherent emotionality (low emotionality, STI; high emotionality, LTI) were either daily exposed to URNS or undisturbed from 17 to 40 days of age (n=32 for each line, in each group). During this time, quail were submitted twice to a sequential feeding procedure: they were offered a hypocaloric diet (7% less caloric than the normocaloric diet) on odd days and a hypercaloric diet (7% more caloric) on even days, for 8 days; then, they received a normocaloric diet (metabolizable energy=12.56 MJ) for 3 days. This sequential feeding procedure was used to assess anhedonia and diet preferences thanks to choice tests (hypo vs. hypercaloric diets) performed at the end of each period. Short-term (30 min) and daily intake were also measured each day. Behavioral tests were performed to assess quail's emotional reactivity. Results showed that URNS enhanced quails' emotional reactivity, e.g. in the reactivity to human test, disturbed quail came later (P=0.011) and spent less time (P<0.01) in front of the cage (close to the human) than control quail. URNS did not change quails' relative preferences for the hypercaloric diet (P>0.1), but URNS reduced their daily intake during the 2nd period (P<0.05). Motivation for each diet (assessed by their short-term intake) was differently affected by URNS during the 2nd period: STI quail decreased their motivation to eat the hypercaloric diet (P<0.01) whereas LTI increased their motivation to eat the hypocaloric diet (P<0.01). In conclusion, both lines of quail experienced a chronic stress as URNS induced an increase of their emotional reactivity. Interestingly, URNS induced opposite changes in quail's feeding behavior: LTI disturbed quail seemed to express a short-term compensatory behavior because of their high motivation to eat, whereas STI disturbed quail seemed to be in a devaluation process as shown by their anhedonia and their decrease of daily intake.

To go or not to go: is that the question – comparison of a go/no-go and active choice design to assess cognitive bias in horses

Sara Hintze[1,2], Emma Roth[3], Iris Bachmann[1] and Hanno Würbel[2]
[1]Agroscope, Swiss National Stud Farm, Les Longs-Prés, 1580 Avenches, Switzerland, [2]University of Bern, Animal Welfare, Länggasstrasse 120, 3012 Bern, Switzerland, [3]AgroParisTech, 16 Rue Claude Bernard, 75005 Paris, France; sara.hintze@vetsuisse.unibe.ch

Cognitive bias tests to assess the valence of animals' emotional states have been developed for several species. In horses, spatial judgement bias tests have been used but it is unclear whether the latency measurements provide valid measures of emotional valence. The aim of this study was to compare two test paradigms based on choices, a Go/No-go Design (GNgD) and an Active Choice Design (ACD), in terms of time needed for training and variation in test outcomes. Five stallions and five mares were randomly assigned to the two designs and trained for 10 trials/day to discriminate between a low and a high tone. In the GNgD, horses learned to approach a bucket for a food reward when presented with one tone, and to stay in the start box when presented with the other. In the ACD, two buckets were presented and the horses were trained to approach the bucket on one side in response to one tone to obtain a big reward and to approach the bucket on the other side in response to the other tone to obtain a small reward. After reaching a predetermined learning criterion, horses were tested for four days with three intermediate tone-trials interspersed between 10 high and low tone-trials. All five GNgD horses but only one ACD horse reached criterion in an average of 38 training sessions (ranging from 25 to 52 sessions). Training was stopped after 40 sessions in the remaining ACD horses whose performance did not improve. During testing, the GNgD horses showed an intermediate response with a slightly positive bias (63%) to the three intermediate tones without discriminating between them. Moreover, individual horses responded differently to the intermediate tones. Thus, horses can be successfully trained on a GNgD based on auditory cues, while a comparable ACD may be too difficult for them to learn. However, further refinements will be needed to render this GNgD a valuable tool for assessing emotional valence in horses.

A novel method to study blanket preferences in horses

Cecilie M Mejdell[1], Turid Buvik[2], Grete H. M. Jørgensen[3] and Knut E. Bøe[4]
[1]*Norwegian Veterinary Institute, P.O.Box 750 Sentrum, 0106 Oslo, Norway,* [2]*Trondheim hundeskole, L. Jenssensgt 47, 7045 Trondheim, Norway,* [3]*Bioforsk Nord, P.O. Box 34, 8860 Tjøtta, Norway,* [4]*Norwegian University for Life Sciences, P.O. Box 5003, 1432 Ås, Norway; cecilie.mejdell@vetinst.no*

Blanketing of horses is a common management procedure. A blanket protects the horse from being wet and cold, but is even used indoors and during summer. We wanted to investigate the horses' own preferences for blanketing. To do so, we had professional trainers to teach horses to use symbols to communicate their wishes. Sixteen adult horses of various breeds, on two farms, went through a 10 step training program. Horses were trained for 2 (3) bouts of 5 minutes each per day, 5-7 days a week. First, horses were trained to approach and touch a board (35×35 cm) using operant reward based conditioning. Thereafter, horses learned by association the difference between the symbol for 'blanket on' and 'blanket off'. Plasticity, i.e. independency of context and signal position was trained, and level of understanding was controlled. At this stage, only meaningful choices were rewarded. When the horse had chosen without error for the last 12 trials, a third symbol, meaning 'stay as is' was introduced and learned. The most critical point was the transition to the free choice situation. From now on, any choice made by the horse was rewarded. This final step started by temperature challenge testing. The horse had to understand that his choice, touching a specific signal, decided the action of the handler regarding blanketing. The horse should even have an idea of the consequence of the choice in terms of thermal comfort the next hours. Speed of learning varied among the individuals, but all 16 horses passed the training program successfully in 14 days. The horses' preferences were successively tested under differing weather conditions with air temperatures from -15 °C to + 20 °C. Some preliminary results are given. Horses were tested with or without blanket in the test situation according to the owner's ordinary management. Horses were left outdoors for two hours before given the choice; either to stay as is or to change blanket status (signals 'stay as is' and 'blanket on' or 'blanket off', dependent on if the horse wore a blanket or not). We found clear individual preferences. In general, cold blooded horses more often preferred to stay without a blanket and warm blooded more often preferred to have a blanket on. All individuals chose to stay outdoors without blanket at some occasions. In rainy, windy or very cold (-15 °C) weather, more horses preferred to have a blanket on compared to days without precipitation or wind. Communication by the use of visual symbols is a promising tool and its use could be expanded to other management practices.

Assessment of human-animal relationship in dairy buffaloes

Giuseppe De Rosa[1], Fernando Grasso[1], Felicia Masucci[1], Andrea Bragaglio[2], Corrado Pacelli[2] and Fabio Napolitano[2]
[1]*Università di Napoli Federico II, Dipartimento di Agraria, Via Università 133, 80055 Portici, Italy,* [2] *Università della Basilicata, Scuola di Scienze Agrarie, Forestali, Alimentari ed Ambientali,, Via dell'Ateneo Lucano 10, 85100 Potenza, Italy; giderosa@unina.it*

The present study was aimed to assess the relationship among stockperson behaviour and buffalo behaviour. The research was carried out in 17 buffalo farms located in southern Italy. Observations were conducted by two trained assessors. Human–animal relationship was assessed performing two different tests: behavioural observations of stockperson and animals during milking and avoidance distance at the manger. These tests were repeated within one month to assess test–retest reliability. Stockpeople attitude was also evaluated using a questionnaire divided into four sections (general beliefs about buffaloes, general beliefs about working with buffaloes, behavioural intentions with respect to interacting with buffaloes and job satisfaction) including 21 statements. A high degree of test–retest reliability was observed for all the variables concerning the behaviour of stockperson and animals. The values of coefficients (Spearman rank correlation coefficient: r_s) ranged from 0.587 (P<0.01) for the number of steps performed by the animals during milking to 0.943 (P<0.0001) for the percentage of animals that can be touched at the manger. A negative correlation between job satisfaction of stockpeople and number of steps performed by the animals during milking was found (r_s=-0.508, P<0.05). The lack of further significant correlations between the questionnaire scores and the behaviour of stockpeople and animals may be due to low number of statements included in the questionnaire or to false answers given by stockpeople. Positive stockperson interactions both in absolute number and in percentage terms negatively correlated with the number of kicks during milking (r_s=-0.564 and r_s=-0.615, P<0.01; respectively) and the percentage of animals treated with oxytocin (r_s=-0.529 and r_s=-0.671, P<0.01; respectively). Conversely, negative stockperson interactions both in absolute number and in percentage terms were positively correlated with the number of kicks during milking (r_s=0.676 and r_s=0.585, P<0.01; respectively) and mean avoidance distance at manger (r_s=0.525 and r_s=0.507, P<0.05; respectively). The present study showed that test-retest reliability of the variables used to assess human-animal relationship in buffalo was high. In addition, distance at manger is suitable for the assessment of the quality of human-animal relationship in buffalo as it is less time consuming than the observation of the behaviour performed by stockpeople and animals during milking.

Hair cortisol and acute phase proteins as stress markers in pigs

Raquel Peña[1], Yolanda Saco[1], Raquel Pato[1], Laura Arroyo[1], Ricard Carreras[2], Antonio Velarde[2] and Anna Bassols[1]
[1]*Universitat Autònoma de Barcelona, Dept. Bioquimica i Biologia Molecular, Fac. Veterinària, 08193 Cerdanyola del Vallès, Spain,* [2]*IRTA, Veïnat de Sies s/n, 17121 Monells, Spain; anna.bassols@uab.cat*

It has been proposed that cortisol accumulation in hair provides a retrospective measure of the HPA axis activity. For this reason, hair may be used as a convenient, non-invasive sample to evaluate chronic stress. On the other hand, stress alone can induce an acute phase response. The aim of this study was to determine hair and serum cortisol, and serum acute phase proteins (APP) in four groups of pigs to assess the effect of sex (male/female, non castrated) and the halothane genotype (free/carrier) on cortisol accumulation and APP levels. A total of 48 pigs (3-5 month old) in 4 groups of 12 animals according to sex and halothane genotype were included in the study. Blood and hair samples were collected once a month throughout a three month period. During this time, all animals were trained to learn to discriminate positive and negative spacial cues between a bucket with or without access to chopped apples according to its position (left or right). Afterwards, each animal was subjected to a test to evaluate the cognitive bias (CB), where the bucket was placed on a central position. CB is related to the pig's emotional state and it has been suggested to influence the ability of the individual to cope with stress. Serum was used to measure cortisol and APPs (haptoglobin, Pig-MAP and CRP) and hair samples to measure cortisol. Haptoglobin was quantified by a haemoglobin binding assay, Pig-MAP and serum cortisol were assessed with a commercial EIA kit and CRP using an immunoturbidimetric method. Hair was extracted with methanol and cortisol was determined by EIA using a salivary cortisol EIA kit, previously validated in our laboratory for pig hair (CVintra: 8.45% at 0.28 µg/dl and 3.75% at 0.99 µg/dl; recovery: 80%). Male had higher Pig-MAP and lower CRP concentration than females (P<0.05). No differences were found due to the CB or to the halothane genotype for any APP. Our results showed a decrease in all three APPs throughout the three month period. The decrease in APP concentration indicates that training manipulation is not stressful for the animals and/or an improved health status in the farm. Male had higher cortisol in serum than females (P<0.05). Cortisol accumulation in hair was observed throughout the time in all groups, reaching higher values in males. No correlation was found between serum and hair cortisol, indicating that both parameters have different physiological meanings. No differences were found due to CB or to the halothane genotype. Our results indicate that this method is sensitive enough to detect an increase in hair cortisol due to a prolonged time of accumulation.

Psychological stress caused by movement restriction affects sucrose detection in nursery pigs

Jaime Figueroa[1,2], David Solà-Oriol[2], José Francisco Pérez[2] and Xavier Manteca[2]
[1]*Universidad de Chile, Animal Production, Av. Santa Rosa 11735, La Pintana, 8820000 Santiago, Chile,* [2]*Universitat Autònoma de Barcelona, Departament de Ciència Animal i dels Aliments, Campus UAB, 08193 Bellaterra, Spain, Spain; figuejaime@gmail.com*

Reward-related deficits experienced by chronic stress involve loss of pleasure commonly known as anhedonia. This study evaluated if pigs could change their ability to prefer sweet solutions due to psychological stress. A total of 240 pigs (42 d old) allocated in 24 pens (10 pigs/pen) were randomly distributed into two groups: SG (12 pens, stressed group) and CG (12 pens, control group). Two animals in each SG pen were randomly selected and subjected to a psychological stress protocol. Selected animals were immobilized for a 3-min period, 3 times a day for 3 consecutive days. The immobilization was performed by placing the pigs into an elevated plastic box with four openings where the pig legs were placed so that the animal was totally immobilized. The ability to prefer sucrose in both groups were measured on two consecutive testing days by performing a 30 minute choice test to SG pigs and 2 pigs randomly chosen in each CG pen, between sucrose (0.5 or 1%) and tap-water after 15 minutes of the last stress session. Sucrose concentrations (0.5 or 1%) and positions (left/right) were counterbalanced across pens and between testing days. Solution intake during the choice test was analyzed with ANOVA by using mixed linear models with the MIXED procedure of the statistical package SAS°. Higher intakes of 0.5% and 1% sucrose solutions over water were observed by CG pairs (51 vs. 10 ml; P=0.015 and 99 vs. 48 ml; P=0.013, respectively). SG pigs did not show different solutions intakes of sucrose 0.5% and water (30 vs. 40 ml). However, at higher concentration (1%) they clearly preferred sucrose (130 vs. 58 ml; P=0.027). Psychological stress in nursery pigs changed their ability to prefer palatable solutions probably because it creates deficits in their hedonic capacity. However, preference for intense sweetness could be more pronounced in stressed animals due to their rewarding deficit.

Using scalar products to refine the interpretative value of an orientation choice test

Nicolas Meunier and Birte L Nielsen

INRA, UR1197 NOeMI, Phase, Bât 325, 78352 Jouy en Josas, France; birte.nielsen@jouy.inra.fr

Choice tests are often used to examine animal preferences. We have developed a new method to evaluate the behaviour in a choice test based on the orientation of the animal. Using rat pups (n=44) in an open field maze with a choice of two odours (maternal faecal pellets vs. clean wood shavings), we obtained frame-by-frame x,y-coordinates of two markers (on the head and on the lower body along the dorsal mid-line) using a video-tracking freeware (Kinovea. org). Two types of vectors were calculated: an animal orientation vector (AOV; body marker to head marker) and a perfect orientation vector (POV; body marker to odour source) for each odour. The angle between the two vectors in each frame was converted into a scalar product ranging from 1 (pup oriented directly towards the odour source) to -1 (facing the opposite direction). Frame-by-frame scalar products were calculated for each odour source, with the mean difference between the two scalar products indicating degree of preference for an odour. The information provided by the mean scalar product difference (MSPD) could not be obtained from other measures, such as velocity (regression Anova: R^2=1.4%; $F_{2,41}$=0.29; P=0.750) or distance moved (R^2=1.7%; $F_{2,41}$=0.36; P=0.699). In a second test (n=16) using faecal pellets from female rats in oestrus and di-oestrus as odour sources, 4 rat pups chose estrus odor and 12 pups did not make a choice (MSPD mean±se: 0.29±0.075; T=3.89; P=0.001). However, including only the 12 non-choosing pups in the MSPD analysis still resulted in a significant preference for estrus odor (mean±SE: 0.16±0.056; T=2.81; P=0.017), which was not the case for other methods. The MSPD thus performs better than methods based on categorical choices by revealing latent preferences even when no direct choice is made. The MSPD provides a new and continuous variable to describe an animal's preference in a choice test. It can be used instead of or in conjunction with existing methods, and allows a better description of the response of individual animals in a choice test situation. This variable is more effective in differentiating animals, thus allowing a reduction in the number of animals or tests necessary to reach significance.

The unit of measurement matters: an example from observations of maternal care in C57BL/6 mice

Jeremy D. Bailoo[1], Xavier J. Garza[2], Richard L. Jordan[2] and George F. Michel[2]
[1]Veterinary Public Health Institute, Division of Animal Welfare, Länggassstrasse 120, 3012 Bern, Switzerland, [2]The University of North Carolina at Greensboro, Department of Psychology, P.O. Box 26170, Greensboro, NC 27402, USA; jeremy.bailoo@vetsuisse.unibe.ch

C57BL/6 offspring were reared under two conditions during postnatal day (PND) 2-14: Early Handling (EH, daily 15 min separation from the dam) and an Animal-Facility Reared (AFR) control group. Maternal behaviour was recorded at four hours and at one hour before 'lights on' on every other consecutive day, beginning on PND 2. For EH dams, reunion with the pups coincided with the latter observation period. Comparisons of frequencies of bouts of nest attendance, licking, quiescent nursing, and activity not licking, versus durations of these behaviours were assessed in these two groups using hierarchical linear modelling. Subjects spent on average 1,550.5 seconds in the natal nest (nest attendance) with a rate of change of -120.75 seconds for every other consecutive day of observation ($P<0.05$); with no differences between our treatment groups being observed. Conversely, when frequencies were our unit of measurement, EH groups had a greater frequency of bouts of nest attendance (M=5.8, SE=2.6; $P<0.05$). This difference was associated with reunion behaviour of the dam with the pups for the EH group (M=12.6, SE=5.1; $P<0.05$). When looking at specific aspects of maternal behaviour, similar patterns were observed regardless of treatment condition. For durations of behaviour, subjects spent on average: (1) 675.13 seconds active but not licking with a rate of change of -66.45 seconds; (2) 87.55 seconds licking with a rate of change of 7.20 seconds; and (3) 664.64 seconds quiescent nursing with a rate of change of -49.16 seconds; for every other consecutive day of observation ($P<0.05$). For frequencies of bouts of behaviour, subjects spent on average: (1) 18.85 bouts of active but not licking with a rate of change of -1.14 bouts; (2) 4.53 bouts of licking with a rate of change of 0.35 bouts; and (3) 7.15 bouts of quiescent nursing with a rate of change of -0.53 bouts; for every other consecutive day of observation ($P<0.05$). As specific aspects of maternal behaviour (licking, quiescent nursing, and activity not licking) can only occur when the dam is in the natal nest (nest attendance), these results suggest that durations of behaviour rather than frequencies of bouts of behaviours are less idiosyncratic and may be more sensitive to the variations in maternal behaviour that emerge as a consequence of brief periods of maternal separation.

Qualitative Behavioural Assessment of intensive and extensive goat farms

Lilia Grosso[1], Monica Battini[1], Françoise Wemelsfelder[2], Sara Barbieri[1], Michela Minero[1], Emanuela Dalla Costa[1] and Silvana Mattiello[1]
[1]Università degli Studi di Milano, DIVET, Via Celoria 10, 20133 Milan, Italy, [2]SRUC, Roslin Institute Building, Easter Bush, United Kingdom; lilia.grosso@unimi.it

Qualitative Behaviour Assessment (QBA) is a whole-animal approach, integrating perceived details of animals' expressive demeanour, using terms such as tense, anxious, or relaxed. To evaluate the validity and repeatability of QBA for dairy goats, two observers assessed 16 goat farms at the same time, using a list of 16 QBA terms based on literature study and discussion with an experienced focus group. There were 8 'housed' (H) farms, where animals were observed in free stall pens with permanent straw litter, and 8 'pasture' (P) farms, where animals were observed in open pasture ranges. One H farm was removed from analysis due to procedural error. QBA scores generated by observers for the 15 farms were analysed together using Principal Component Analysis (correlation matrix, no rotation). Observer agreement for farm scores on PCA Components (PCs) and on separate QBA terms was investigated using Pearson and Spearman correlations respectively. The effects of housing system and observer on PC scores were analysed using analysis of variance (treatments=observer, housing system and their interaction; block=farm). PCA distinguished three meaningful dimensions of goat expression: PC1 (29%) 'content/calm-frustrated/aggressive'; PC2 (20%) 'curious/attentive-calm/bored; PC3 (12%) 'sociable/playful-alert/agitated'. Farm scores generated by the two observers on the three PCs were significantly correlated (PC1: r=0.75, P<0.001; PC2: r=0.67, P=0.006; PC3: r=0.69, P=0.004). Observers' farm scores on separate QBA terms were significantly correlated for 7 out of 16 terms (P<0.05), and approached significant correlation for an additional 2 terms (P<0.1), indicating an integrated PCA approach to QBA to be more robust. There were significant effects of housing system on PC1 (P=0.05) and PC2 (P=0.02), indicating goats on P farms to be more 'content/calm', and more 'curious/attentive', than goats on H farms. There was a significant observer effect on PC2 (P=0.04), and a significant observer by housing interaction on PC3 (P=0.009). These results show good inter-observer reliability across three dimensions of goat demeanour. However observers differed in their quantification of several QBA terms, indicating the need for further training and refinement of the descriptor list. QBA found the goats' demeanour on H and P farms to differ along two dimensions, suggesting that access to pasture may have a positive effect on goats' emotional state. In sum, these results suggest that, given further refinement, QBA could make a valuable contribution to goat welfare assessment protocols.

Cortisol concentrations and infectious disease in Swedish shelter cats

Elin N Hirsch[1], Eva Hydbring-Sandberg[2], Jenny Loberg[1] and Maria Andersson[1]
[1]Swedish University of Agricultural Sciences, Department of Animal Environment and Health, Box 234, 532 23 Skara, Sweden, [2]Swedish University of Agricultural Sciences, Department of Anatomy, Physiology and Biochemistry, Box 7011, 750 07 Uppsala, Sweden; elin.hirsch@slu.se

Cortisol can be measured in several media and is commonly used to measure physiological stress. Short term increases in cortisol concentration enhance the immune system, but inhibits immunity when prolonged. Here we evaluate non-invasive salivary cortisol measurement as an alternative for invasive plasma cortisol in cats, and investigate potential correlations between cortisol levels, group and shelter size on occurrence of infectious disease. Eleven Swedish shelters were visited and maximum 10 cats per shelter participated in collection of saliva and plasma (cortisol), swab from the conjunctiva (feline herpesvirus [FHV], Chlamydophila felis and Mycoplasma felis) and mouth-swab (metagenomic analysis; reported elsewhere). Eighty-nine cats participated. Eleven saliva samples yielded enough amounts for individual cortisol analysis, the rest were pooled per shelter before analysis. Individual plasma samples were analysed for 83 cats, mean (+SD) 160+132.9 nmol/l, of which 10 had individual salivary cortisol concentrations analysed, mean (+SD) 4.7+2.5 nmol/l. There was no correlation between salivary and plasma cortisol levels, tested with Pearson correlation. Plasma cortisol concentrations were negatively correlated with number of cats per shelter (r_p=0.222; P=0.043). The eye-swab resulted in 2 FHV and 3 C felis positive cats. Due to the low numbers, no statistical analysis was performed. Low positive outcome from the eye-swabs could have been caused by the assay technique (multi-PCR). Lack of correlation between saliva and plasma cortisol might have been caused by several factors e.g. confounding factors at the different shelters, stress effects due to handling, blood contamination of saliva samples, small amount of saliva, and low number of participating individuals. In conclusion, cortisol sampling from plasma and saliva was difficult to perform in cats; handling during sample collection can affect cortisol levels per se. Despite this, there were correlations between cortisol concentration, and shelter size which is interesting to investigate further.

Effect of space on social behaviour in group-housed cats

Jenny Loberg and Frida Lundmark
Swedish University of Agricultural Science, Animal Environment and Health, P.O. Box 234, 532
23 Skara, Sweden; jenny.loberg@slu.se

The ancestor of our domestic cat, the African wild cat, is a solitary species. However, during domestication the cat has become more tolerant towards conspecifics and can function in groups in many contexts. In Sweden, the regulation has just changed from 15 adult cats on 15 m^2 to 15 adult cats on 30 m^2. Little is known how the cats can deal with this social context. The aim of this study was to observe the social interactions in six groups of 15 adult cats housed in 1 m^2/cat, 2 m^2/cat and 4 m^2/cat. Each of the six groups of cats was observed in all three treatments in a balanced order. All groups were mixed sexes and all cats were neutered. After two weeks of acclimatization, three days of behavioural observations were performed, 2 hours in the morning just after feeding and 2 hours in the afternoon just before feeding. Social interactions (aggressive and non-aggressive behaviours), use of resources and cat-stress-score were recorded. In this presentation only results from the social interactions will be reported. The effect of treatment on social interactions was analysed using a non-parametric Kruskal-Wallis test. There were no significant differences between the treatments in the frequency of aggressive movements per 2 h observation either before feeding (P=0.96, Median 1 m^2=40, 2 m^2=24, 4 m^2=32.5) or after feeding (P=0.96, Median 1 m^2=18.3, 2 m^2=21.2, 4 m^2=24.7). There were many non-aggressive vocalisations before feeding but no difference were found between treatments (P=0.27, Median freq/2 hour 1 m^2=133, 2 m^2=141, 4 m^2=221). In this data of only six groups of cats, we could not see a clear difference in the amount and the type of social interactions between group-housed cats. In conclusion the increase in available space did not change the social behaviours. But it is questionable if this is a tolerable amount of aggressive behaviours in a group of cats or if it compromises their welfare.

Effect of environmental enrichment type on the behavioural and glucocorticoid responses of bold and shy cats to single caging

Jacklyn J. Ellis, Henrik Stryhn, Jonathan Spears and Michael S. Cockram
Atlantic Veterinary College, University of Prince Edward Island, Sir James Dunn Animal Welfare Centre, Department of Health Management, 550 University Ave, Charlottetown, C1A 4P3, Canada; jjellis@upei.ca

This study investigated whether environmental enrichment (EE) would reduce the stress responses of singly housed cats, and if either the type of EE or whether the cat was bold or shy would influence this relationship. Seventy-two cats were housed for 10 days with either a hiding box (BOX), perching shelf (SHELF), or no additional EE (CTRL). Using an emergence test from a cat carrier with a cut-point of 10 s to differentiate behavioural style, bold and shy cats were balanced between groups. Continuous observations for two 4-h periods/day/cat of activity, location, and posture were conducted from video-recordings. Daily measurements were made of food intake, Cat-Stress-Score (CSS), and faecal glucocorticoid metabolite (FGM) concentration. Dichotomous CSSs (<3 or ≥3) were analysed using a generalised estimating equation. Other variables were analysed using a linear mixed model with a repeated measures correlation structure and fixed effects of day, treatment group, behavioural style, and all possible interactions. Cats in BOX had lower FGM (BOX=126 ng/g, CTRL=436 ng/g, P=0.02), and consumed more food (BOX=62 g/d, CTRL=39 g/d, P=0.03), than cats in CTRL. Shy cats had a greater probability of registering a CSS≥3 (shy=0.37, bold=0.02, P<0.01), had a greater CSS on days 1-3 (P<0.05), and within the treatment group BOX, and spent a greater percentage of time in the hiding box (shy=87%, bold=75%, P=0.03) than did bold cats. Cats in the SHELF group only spent a median of 1% of their day on the shelf. Food intake and percentage time spent eating increased, and percentage of time spent grooming decreased across time. Results indicated that the stress of single caging can be dampened by including a hiding box. The absence of a significant interaction between treatment group and bold/shy behavioural style suggests that a hiding box was beneficial regardless of whether a cat was bold or shy.

Behavior of community dogs in two cities in Southern Brazil

Larissa Helena Ersching Rüncos[1], Gisele Sprea[2], Edson Ferraz Evaristo De Paula[3] and Carla Forte Maiolino Molento[1]
[1]*Universidade Federal do Paraná, Rua dos Funcionários, 1540, 80035050, Brazil,* [2]*Prefeitura de Campo Largo, Avenida Padre Natal Pigatto, 925, 80.040-280, Brazil,* [3]*Prefeitura de Curitiba, Rua Presidente Faria, s/n, 80020290, Brazil; lari.hr@gmail.com*

Some cities in Brazil are identifying community dogs and turning their condition official in a partnership between municipality and local community. Once dogs are vaccinated and receive medical attention, the risks of disease transmission are minimized. However, they can exhibit behaviors that may be considered nuisance. Our objective was to study and compare the behavior of community dogs in two cities in Southern Brazil. A total of 105 dogs were studied, 73 in Campo Largo and 32 in Curitiba, which corresponds to all community dogs registered in both cities. The behavior was studied through a questionnaire with caretakers and direct dog observation. The majority of dogs slept during the day (47.6%, 50/105); ran frequently (81.9%, 86/105); and played frequently (99.0%, 104/105). The minority of dogs was seen far from their living places (14.3%, 15/105). Community dogs in Campo Largo barked more frequently ($P<0.01$) and ran more frequently ($P<0.05$) than dogs in Curitiba. Display of territorial behavior was different ($P<0.05$): a higher proportion of dogs did not allow unknown dogs approach them in Curitiba (31.3%, 10/32) than in Campo Largo (15.1%, 11/73). Some dogs (18%, 19/105) had a history of biting aggression to other dogs. Of the dogs, 49% (52/105) jumped to play with their caretakers and 45.7% (48/105) displayed friendly behavior towards unknown people in the streets; 68% (72/105) chased vehicles, most frequently motorcycles. Bites toward people were described in a higher proportion ($P<0.05$) of dogs in Curitiba (9.4%, 3/32) than in Campo Largo(6.8%, 5/73). All behaviors that did differ between cities may be related to the environment in which the dogs were living. Community dogs in Campo Largo lived in neighborhood streets and in Curitiba lived in bus stations, which may be perceived by the dogs as a more delimited territory, and are characterized by greater people flow. Community dogs displayed mostly docile behavior and positive interactions with their caretakers, other people and other dogs. They can express most of normal dog behavior, which is positive for their mental health. However, some dogs presented behavioral problems, and it is necessary to study with more detail the problematic events in order to reach behavioral diagnosis and seek solutions, perhaps through behavioral modification. The viability of permanence of some of the dogs as community animals must also be considered. Funded by Fundação Araucária, CAPES, and Campo Largo Town Hall.

Evaluating dog behaviour in everyday life

Lina Roth and Per Jensen
Linköping University, The Department of Physics, Biology and Chemistry, Avian Behavioural Genomics and Physiology Group, 58183 Linköping, Sweden; linaroth@ifm.liu.se

The dog is our oldest domestic animal and during thousands of years of coexistence a close relationship between dogs and humans has evolved. Today the dog is seen upon as our best friend and a number of extensive and standardized tests and questionnaires are used to enhance our understanding of dog behaviour. Our aim was to develop and validate a simple test utilising a few minutes of video recordings in an everyday situation to characterise behavioural variations in dogs. We visited 14 dog courses targeting companion dogs and their owners, and made video recordings from 86 dogs of a wide variety of breeds. The owners were asked to walk up to a novel object with their dog and fill in a short questionnaire. The main purpose of the survey was to keep the owners occupied for a couple of minutes while their dogs, without commands, were free to do whatever they liked. The survey asked the owners to grade (on a scale of 1-5) whether they perceived the dog as stressed, cooperative, aggressive towards dogs or humans, happy to meet dogs or humans, curious and playful. During the survey, three minutes of video recording was obtained from each dog and later analysed with 1/0 sampling (5 second intervals). Behaviours including contact seeking behaviour with both the owner and a stranger, general activity, and interaction with the novel object were recorded. Using Spearman's nonparametric correlation we found a significant positive correlation between the owner's grading of how cooperative his or her dog is and the frequency of the dog seeking eye contact with its owner ($r=0.36$; $P<0.01$). Dogs with high cooperation score also looked back at their owner more often after having met a stranger than dogs with low score ($r=0.39$; $P<0.01$). In addition, a negative correlation between the cooperation grade and leash pulling was found ($r=-0.32$; $P<0.01$). Hence, dogs that the owner perceived as cooperative sought more often eye contact and pulled the leash less often than other dogs. Furthermore, females jumped significantly more on their owners than males and male dogs pulled the leash more than females (Mann-Whitney test, $P<0.05$). The results suggest that it is possible to identify at least some stable behavioural characteristics by studying a dog for only a couple of minutes in an everyday situation. Hence, this simple test can be used for obtaining behavioural data from large populations of dogs for, e.g. breed comparisons and genetic analysis.

Comparison of welfare consequences and efficacy of pet dog training using electronic collars and reward based training

Jonathan Cooper, Nina Cracknell, Jessica Hardiman, Hannah Wright and Daniel Mills
University of Lincoln, LifeSciences, Riseholme Campus, Lincoln LN2 2LG, United Kingdom;
jcooper@lincoln.ac.uk

The use of electronic collars is controversial with opponents suggesting they are cruel and barbaric and have no place in dog training, nevertheless no large-scale studies have been conducted on their use in the pet population. This study investigated the performance and welfare consequences of training dogs in the field with manually operated electronic devices (e-collars). Following a preliminary study on 9 dogs, 63 pet dogs referred for poor recall and related problems were assigned to one of three Groups: Treatment GroupA were trained by industry approved trainers using e-collars, vocal commands and a collar mounted vibration warning cue; Control GroupB trained by the same trainers but without use of e-collars; and GroupC trained by Association of Pet Dog Trainers again without e-collar stimulation (n=21 for each group). Dogs received two 15 minute training sessions per day for 4-5 days. All studies were undertaken following local (UK) ethical approval. Training sessions were recorded on video for behavioural analysis. Saliva and urine were collected to assay for cortisol over the training period. During preliminary studies there were negative changes in dogs' behaviour on application of electric stimuli, and elevated cortisol post-stimulation. These dogs generally experienced high intensity stimuli without pre-warning cues during training. In contrast, in the subsequent larger, controlled study, trainers used lower settings with a pre-warning function and behavioural responses were less marked. Nevertheless, GroupA dogs spent significantly more time tense, yawned more often and engaged in less environmental interaction than GroupC dogs. There was no difference in urinary corticosteroids between groups. Salivary cortisol in GroupA dogs was not significantly different from that in GroupB or GroupC, though GroupC dogs showed higher measures than GroupB before and after training. Following training 92% of owners reported improvements in their dog's referred behaviour, and there was no significant difference in reported efficacy across groups. Owners of dogs trained using e-collars were less confident of applying the training approach demonstrated. These findings indicate that there is no consistent benefit to be gained from e-collar training but greater welfare concerns compared with positive reward based training

Behaviour and experiences of dogs during the first year of life predict the outcome in a later temperament test

Pernilla Foyer[1,2], Nathalie Bjällerhag[2], Erik Wilsson[3] and Per Jensen[2]
[1]*Swedish National Defence College, Department of Military Studies, Military-Technology Division, Swedish National Defence College, 115 93 Stockholm, Sweden,* [2]*Linköping University, IFM Biology, Behaviour Genomics and Physiology group, Linköping University, 581 83 Linköping, Sweden,* [3]*Swedish Armed Forces Dog Instruction Centre, FHTE, 195 24 Märsta, Sweden; pernilla.foyer@fhs.se*

Experiences early in life are known to shape the behaviour development of animals, and therefore events occurring during the preadolescence and adolescence may have long-term effects. In dogs, this time period may be important for their later behaviour and thereby also their suitability for different working tasks. In this study we used the breeding practice for Swedish military working dogs to investigate this possibility. German Shepherds were bred at a central facility and then kept in host families for about a year, before participating in a standardised temperament test determining their suitability for further training. We surveyed the link between the behaviour of 71 prospective military working dogs in their home environment during the first year of life, as assessed by a modified and amended C-BARQ survey, and the performance of the dogs in a temperament test (T-test) applied at about 15 months of age. To investigate whether there were any relationships between the behaviour in the T-test and the C-BARQ categories, Pearson correlation coefficients between success rate in the T-test (Index Value) and C-BARQ categories and items were calculated. Dogs with higher success rate in the T-test scored significantly higher for C-BARQ category Trainability ($t_{(69)}$=3.82, P<0.001) and C-BARQ items Hyperactivity/Restlessness, Difficulties in Settling Down ($t_{(69)}$=2.25, P=0.028), and Chasing/Following Shadows or Light Spots ($t_{(68)}$=2.16, P=0.035). Furthermore were dogs with a higher success rate in the T-test left significantly longer alone at home per day (2.97±0.32 vs. 2.04±0.33 h/day; $t_{(66)}$=2.00, P=0.050). Dogs with a lower success rate in the T-test showed significantly higher score for the C-BARQ categories Stranger-Directed Fear ($t_{(34)}$=-2.48, P=0.018), Non-social Fear ($t_{(43)}$=-3.01, P=0.004), and Dog-Directed Fear ($t_{(29)}$=-2.38, P=0.024). The results indicate that the experiences and behaviour of the dogs during their first year of life is important in determining their later behaviour and temperament, something that could potentially be used to improve selection procedures for working dogs. An unsuspected result was that success in the T-test was correlated to behaviours usually associated with problem behaviour, which calls for a deeper analysis of the selection criteria used for working dogs.

Behavioral problems in dogs: analysis of cases in an Australian clinic

Ramazan Col[1,2], Cam Day[3] and Clive Phillips[1]
[1]*School of Veterinary Science, University of Queensland, Centre for Animal Welfare and Ethics, Centre for Animal Welfare and Ethics, School of Veterinary Science, University of Queensland, 4343 Gatton, Queensland, Australia,* [2]*University of Selcuk, Faculty of Veterinary medicine, Physiology, University of Selcuk, Faculty of Veterinary medicine, department of Physiology, 42075 Konya, Turkey,* [3]*Cam Day Consulting, Cam Day Consulting, Brisbane, QLD 4102, 8406, Australia; rcol@selcuk.edu.tr*

We analysed behavior problems in dogs (n=7858) presented at a Brisbane companion animal behaviour clinic between 2001 and 2013. Dog owners completed a registration form indicating their identity, location, the suspected behaviour problem and type of dog: self-assessed breed, purebred or crossbred, sex, neutered/entire, and age. It was completed either on-line or by telephone. The on-line form included 22 multiple-choice questions about the behaviour problems. Breed types, of which there were 50, were also categorized according to the Australian National Kennel Council groups. Data were analysed by descriptive statistics. In total, 34.0% of the dogs were of mixed breeds; 57.8% were purebreeds and 8.2% were of unspecified breed. The main behaviour problems were aggression (2928), followed by Barking (1101), Anxiety (919), Noise Fear (388), House soiling (337), Other fearfulness (313), Destructive behaviour (295), Boisterousness (186), Escape behaviour (178), and Disobedience (146). The breed groups represented were Working (20%), Terriers (19%), Toys (15%), Gundogs (11%), Utility (10%), Non Sporting (10%), Hounds (6%). The most common breeds were Staffordshire Bull Terrier (725), Maltese (493), Border Collie (439), Labrador (366), German Shepherd (361), The sex distribution was 46.9% males (24.6% entire, 75.5% neutered) and 40.5% females (21.4% entire, 78.6% neutered, with 12.7% unknown sex. Awareness of the major behavior problems will alert owners and others to their importance and contribute towards the welfare of dogs.

Social modulation of yawning behavior in the domestic horse – an exploratory analysis

Rachele Malavasi

CNR, National Research Council, IAMC, piazzale Aldo Moro 7, 00185 Rome, Italy; rachele.malavasi@iamc.cnr.it

In social species, yawning behaviour is considered to serve as social cue. In particular, contagious yawning seems to reflect an empathic relation between individuals in human and non-human primates, and probably in dogs. In horses, highly social animals, yawning has been explored only in isolation, with results related to stereotypic behaviours, psychological stress or physical diseases. This is an exploratory analysis of the social modulation of yawning behaviour in the domestic horse. The study was conducted on a herd of 5 males and 3 females, living in semi-free ranging conditions at a rescue center in Italy (IHP). From 10 hours of video-recordings, 89 yawning events were collected: 45 spontaneous events, defined as the condition where A yawns while B is close to A but does not yawn, and 44 events apparently derived from contagion, defined as the condition where B yawns within a 4 minutes time-window from the yawn of an individual A. The two kind of yawning events seem to occur at random (P=0.46, One-way Anova), even between affiliate individuals ('friends'; P=1.00, One-way Anova), suggesting that contagious yawning may be absent in domestic horses. Nevertheless, social interactions (grooming, body contact, aggression) and yawns seem to be reciprocally influenced. Interactions between friends preceding a yawn event induce an increase in the number of yawns during a sequence (P=0.001, vs P=0.29 in non-friends; runs test). After a yawn event, interactions do not increase (P=0.38, One-way Anova), but they are nevertheless more frequent between friends (P=0.001, One-way Anova). The act of yawning seems to affect the immediate social behaviour of the yawner, that switches from a passive to an interacting condition, and vice versa (P=0.03, c-score), or it moves closer or narrower to other individuals (P=0.001, c-score). These preliminary data support a social modulation of yawning behaviour in the domestic horse, that seems related with affiliative relations between subjects. Further analysis are needed to describe this behaviour in domestic horses and to assess its role in the species' social dynamics.

Vertebral and behavioural problems are related in horses: a chiropractic and ethological study

Clémence Lesimple[1], Carole Fureix[1], Hervé Menguy[2] and Martine Hausberger[1]
[1]Université de Rennes1, UMR-CNRS 6552, Laboratoire EthoS, 263 avenue du Général leclerc, 35042 Rennes cedex, France, [2]Chiropractic office, 1 rue Ernest Psichari, St Jacques de la Lande, France; lesimple.c@gmail.com

Although vertebral problems are regularly reported on riding horses, these problems are not always identified nor noticed enough to prevent these horses to be used for work. On another hand, behavioural problems, in particular aggression towards humans, are a common source of accidents in professionals (e.g. the third after dogs and bovids in veterinarians). In the present study, we hypothesized that part of these undesirable reactions may be due to altered welfare, and possibly back pain. Thus, fifty nine horses from 3 riding centres were submitted on one hand to standardized behavioural tests in order to evaluate their reactions to humans, on another hand to chiropractic examination performed by an experienced chiropractor, who was 'blind' to the results of the behavioural tests and did not know the horses beforehand. The results show (1) that about half (51%) of the horses showed at least once an aggressive reaction towards the experimenter, (2) that 73% of the 59 horses were severely affected by vertebral problems, while 23% were mildly or not affected by such problems. Interestingly, horses severely affected were statistically more prone to aggressive reactions towards a human than the others, confirming a possible relation between back pain and aggressiveness. Differences could be observed between riding centres on both aspects (behavioural and vertebral) and we investigated whether differences in working conditions may explain such differences. Nineteen of these horses, corresponding to two riding centres, could be followed at work (riding lesson) using video recording: it appeared that the degree of vertebral problems identified at rest was statistically correlated with the riders' (hands' height, rein length and leg position) and horses's (neck height and curve) attitudes at work. Riders and horses attitudes clearly differed between centres. These promising results are part of a larger scale study on horses' welfare and suggest fruitful new lines of research based on crossing of chiropractics and behavioural studies.

Equine social organization: a new adaptive view

Lucy Rees
Eduquina, km.480, ctra 630, 10600 Plasencia, Spain; lucyrees5@gmail.com

This paper presents a defence-based interpretation of equine social relations, with critical reassessment of the dominance hierarchy paradigm. Horses are prey animals without primary resource competition, characteristics reflected in their social relations. Feral Venezuelan criollos threatened by predators show self-organizing massed flight described by three factors: cohesion, synchrony, and collision avoidance. Although stallions usually initiate flight, there are no fixed leaders. This behavioural algorithm governs coordinated group movements in many species. During maintenance activities feral bands follow the same algorithm but differing individual requirements may produce incomplete synchrony during grazing. Individual motivation initiates changes with which band members synchronize. Synchrony and cohesion appear innate, but collision avoidance (respect for individual space) is taught by in-band aggression, its final cause. Proximal causes, often age-related, vary. In domestic horse social dynamics the flight algorithm is overlaid to different extents by winner/loser effects learned in unnatural focal food competition, combined with aggression resulting from differing individual responses to behavioural stress. Contributory factors include overcrowding, unnatural and unstable groupings, weaning, lack of social education and uncomfortable work. A resultant avoidance order does not reduce aggression These variables produce the conflicting rank correlations typical of dominance hierarchy studies, which furthermore lump in-band aggressions of different proximal causes, make unjustified assumptions about their final cause, falsely parallel feral and domestic horse behaviour and ignore adaptive function in social organization. Despite these theoretical shortcomings the dominance hierarchy paradigm is that promoted by ethologists. The ethological definition of dominance differs from that used in Standard English, producing widespread public misunderstanding detrimental to human-horse relations and training methods. Following the lead of canine ethologists, equine ethologists should move on to repair the damage done to horse welfare by the misleading concept of in-band dominance hierarchies, and present horse social behaviour in the light of its adaptive functions.

Assessing temperament of Murgese foals

Ada Braghieri[1], Giuseppe De Rosa[2], Andrea Bragaglio[1], Giuseppe Malvasi[1] and Fabio Napolitano[1]
[1]Università della Basilicata, Scuola di Scienze Agrarie, Forestali, Alimentari ed Ambientali, Via dell'Ateneo Lucano 10, 85100 Potenza, Italy, [2]Università di Napoli Federico II, Dipartimento di Agraria, Via Università 133, 80055, Italy; ada.braghieri@unibas.it

Twenty-seven Murgese foals (6 males 21 months old, 10 males 9 months old, 11 females 9 months old) were used to assess the effects of age and sex on their response to three different environmental challenges: open field (OF), novel object (NO) and bridge test (BT). Testing order was always the same: OF, NO and BT. Tests were conducted at week-intervals in a 11×10 m outdoor paddock, novel to the animals, with earth floor and solid walls. In the OF and NO (the novel object was a pink ball of 0.41 m diameter) tests, animals were individually confined for 3 and 4 min, respectively, and their behaviour video-recorded. In BT an unfamiliar person led the horse trying to make it cross the bridge (a foam blue mattress of 200×100×10 cm). The test was stopped after 10 min or when the horse crossed the bridge with at least three feet. Younger males as compared with older males showed higher per cent vigilance (20.58±4.23 vs. 0.00±5.47%, t=2.98, P<0.01, and 28.25±4.77 vs. 8.26±6.15%, t=2.57, P<0.05, in OF and NO, respectively), sustained walking in OF (10.98±2.94% vs. 0.00±3.79%, t=2.29, P<0.05, respectively) and number of vocalisations in NO (2.3±0.40 vs. 0.17±0.52, t=3.23, P<0.01, respectively). Younger males as compared to older males also showed higher latency time to touch the ball during the NO (3.23±0.41 vs. 1.82±0.53 min, t=2.10, P<0.05, respectively) and to cross the bridge (8.16±1.19 vs. 4.38±1.54 min, t=2.00, P<0.05, respectively). Older males as compared with younger males were all able to cross the bridge (100 vs. 30%, χ^2=7.47, P<0.01, respectively) and to touch the ball during NO (100 vs. 60%, χ^2=3.20, P<0.10, respectively). At 9 months of age females as compared with males of the same age exhibited lower latency time to touch the ball during NO (1.60±0.43 vs. 3.23±0.46 min, t=2.59, P<0.05, respectively) and tended to show a higher number of touches and time spent nosing the ball (1.73±0.36 vs. 0.80±0.38, t=1.76, P<0.10 and 1.66±0.44 vs. 0.50±0.46%, t=1.81, P<0.10, respectively). In addition, a higher percentage of females crossed the bridge as compared with males of the same age (72.7 vs. 30.0%, χ^2=3.83, P<0.05, respectively). They also vocalised more than the males of the same age both in OF and in NO tests (6.91±1.39 vs. 2.70±1.46, t=2.08 and 6.91±1.45 vs. 2.30±1.53, t=2.18, P<0.05, respectively). Several variables were able to discriminate among different categories of Murgese foals. Further studies are needed to assess consistency across ages and generations for possible inclusion of reliable parameters in future breeding programmes.

Lusitano horse stress and metabolic adaptive responses to different handling procedures

Rosana Galán[1], Juan Manuel Lomillos[1], Ramiro González[2] and Marta E. Alonso[1]
[1]*University of León, Animal Production Department, Facultad de Veterinaria, 24071. León, Spain,* [2]*University of León, Medicine, Surgery and Anatomy Department, Facultad de Veterinaria, 24071 León, Spain; marta.alonso@unileon.es*

The effects of exercise on the welfare and the metabolism in thoroughbreds have been widely studied over the past years from a biochemical and physiological point of view. However, only limited studies have been carried out in Lusitano horses used in bull-fighting. The objective of this study was to assess the stress and the metabolic adaptive responses of Lusitano horses to handling procedures of training, transport and bull-fighting. Twenty seven horses between the ages of 5 and 17 years were selected from 3 different ranches in Spain. Five blood samples were taken from the jugular vein in all horses at 5 different times: resting, post-training, post-transport, and before and after the bull fight. Gasometrical and haematology analyses were assessed using a portable i-STAT analyser, while electrolyte and biochemical analyses were carried out at LTI at the University of Leon using a COBAS INTEGRA 400 and MULTI-ANALITE analyser. Data was analysed using a 2-way ANOVA and a linear mixed model in SAS. Multiple comparisons were adjusted with Tukey's test. Compared to the post-training period, bullfighting significantly increased lactic acid values (1.8 vs. 12.7 mmol/l, t=9.93, P<.0001) but the activation of metabolic pathways responsible for maintaining acid-base balance kept pH within physiological levels in sport horses (7.45 and 7.33, respectively). Differences in cortisol (5.3 and 4.6 μg/dl), heart (57 and 60 bpm) and respiratory (42 and 41 bpm) rates, and body temperature (38.6 and 38.1 °C) were not significant between post-training and post-bullfighting samples. Our results also suggest that the horses studied used for bullfighting are well adapted to transport owing to the absence of significant differences from the values found at rest. In conclusions, (1) transport handling does not represents either any source of stress or requires a metabolic adaptation effort by horses trained for bull-fighting and (2) training exercise on the farm and fighting produced similar stress reactions and Lusitano horses are capable of coping with the metabolic demands of both situations.

Pet exaptation index indicates potential adaptation of mammal species for living in human environments

Paul Koene

Wageningen Livestock Research, Animal Welfare, De Elst 1, 6708 WD Wageningen, the Netherlands; paul.koene@wur.nl

Wild characteristics in animal species may exapt species for domestication. Such characteristics may also give a good indication whether animal species adapt readily to a human environment, show few welfare problems and are suitable as pets. In a project to determine the suitability of mammals the so-called Pet Exaptation Index (PEI) was determined based on the presence or absence of such characteristics. Based on an internet inquiry in which keepers were asked for the species they keep as companion animal in the Netherlands, 90 mammal species were selected. In the follow-up scientists collected citations in literature about species characteristics and filled in the PEI-questionnaire after collecting all information. Another group of experts also filled in the PEI-questionnaire. The dataset included 9 assessments of characteristics per species and averages of favourable and unfavourable characteristics, reliability and PEI were calculated (range 0.249-0.738). Cat species (e.g. the leopard cat and the serval) had a low PEI (0.325, 0.348) and cavies and camels (e.g. capybara and Brazilian cavy, lama and Bactrian camel) had a high PEI (0.704, 0.738, 0.702, 0.651). To investigate whether the PEI had also value on family level, data from families including at least 3 of the 90 species were analysed. Of the 7 families in the analysis the canidae, macropodidae, sciuridae had a low PEI (0.466-0.483) that differed significantly ($F_{6,51}$=4.94, P<0.0001) from the caviidae (0.579) and camelidae (0.643). Mammal species of the procyonidae and muridae were intermediate families (0.526 and 0.595). On a species level the PEI may help in determining which species are exapted and may adapt to the human environment. On a family level the results are partly in disagreement with the results determined in a framework to determine mammals suitable as pets. For instance, macropodidae are more than expected on the PEI kept as companion animals.

Avoidance distance at the feed rack in beef bull farms and associated factors

Marlene K. Kirchner[1], Heike Schulze Westerath[2], Ute Knierim[2], Elena Tessitore[3], Giulio Cozzi[3] and Christoph Winckler[1]
[1]University of Natural Resources and Life Sciences, Vienna (BOKU), G. Mendelstr.33, 1180 Wien, Austria, [2]University of Kassel, Witzenhausen, Germany, [3]University of Padova, Legnaro, Italy; marlene.kirchner@boku.ac.at

The Welfare Quality® (WQ) system uses a multi-criteria approach and is assessing farm animal welfare using animal-based measures. A range of single measures are aggregated to 11 'criteria' and 4 'principle' WQ-scores, using expert-derived thresholds for the welfare judgement. The avoidance distance at the feed rack (ADF) is the single measure used to reflect the criterion of 'Good human-animal relationship (HAR)', which is calculated from the proportions of animals in avoidance distance classes (ADF1: 0 cm, ADF2: 1-50 cm, ADF3: 51-100 cm and ADF4: >100 cm). It was the aim of the present study to investigate relationships between farm factors and WQ-scores for the certain avoidance classes (ADF1-4) and the aggregated criterion score (HAR). Data originating from welfare assessments following the WQ system on 62 beef farms in Austria, Germany and Italy were used. Associations between potential influencing factors and HAR measures as well as WQ-scores were first tested at univariate level; all variables with $P \leq 0.2$ in the univariate analysis were included in the final multivariate analysis. The aggregated criterion score HAR was associated with the 'No. cattle on the farm', 'No. persons working with bulls', 'Fattening months', 'Years of beef fattening as production branch', 'Level of feeding place' and 'Incidence of agonistic behaviour' ($F_{6,52} = 6.738$, $P<0.001$). The percentage of animals that could be touched (ADF1) was associated with 'Country', 'Space allowance per animal', 'Animals with diarrhoea', 'Dirty animals' and 'Lean animals' ($F_{6,55} = 7.099$, $P<0.001$). The percentage of animals with ADF2 (<50 cm) was affected by 'Country', 'Painful management procedures', 'Income from bulls', 'Animals with diarrhoea' and the criterion score for 'Social behaviour' ($F_{6,54} = 6.727$, $P<0.001$), whereas factors influencing ADF3 (50-100 cm) were 'Income from bull fattening', 'Years of beef fattening as production branch', 'Agonistic behaviour' and 'Mortality' ($F_{4,55} = 3.371$, $P=0.015$). Finally, the proportion of very shy animals (ADF4: >100 cm) was influenced by 'No. bulls per stockman', 'No. persons working with bulls', 'Fattening months', 'Years of beef fattening as production branch' and 'Incidence of agonistic behaviour' ($F_{5,53} = 5.985$, $P<0.001$).

Effect of a light and heat source in the creep area on piglets' thermal comfort and use of the heated area

Lene Juul Pedersen, Mona Lilian Vestbjerg Larsen and Karen Thodberg
Aarhus University, Dept. of Animal Science, Blickers Allé 20, 8830 Tjele, Denmark;
lene.juulpedersen@agrsci.dk

High and early use of a heated creep area is considered important to prevent hypothermia of suckling pigs and thus to assure high survival and growth. We hypothesise an earlier and increased use of the creep area if the area is enlightened compared to dark and when using a new radiant heat source with a more even and widespread heat surface (Eheater) compared to an incandescent control lamp with light. We used a total of 40 sows over two replicates, randomly distributed to three Treatments: Control lamp with light (n=20), Eheater without light (n=10) and EHeater with light (n=10). Observations of piglets' use of creep area were made as scan sampling every ten minutes for 3 h during two Periods (Day: period with daylight; Night: period with dark) on day 1, 2, 3, 7, 14, and 21 after farrowing. On the same days the piglets' rectal temperature and weight were measured. The results showed a significant interaction between Treatments and Periods ($F_{2,191}$=3.2; P=0.04) with increased use of the creep area during night if kept dark (Eheater without light) compared to the Control and Eheater with light (Night: 39, 30% and 32%, respectively). The piglets used the creep area more during the Day than during Night (Day: Eheater without light, Control, Eheater with light; 41, 40, 39% of piglets, respectively). The rectal temperature were not influenced by treatments but by a weight and day interaction ($F_{5,2469}$=15; P<0.0001) with a strong positive effect of piglet weight on rectal temperature during the first 7 d but not hereafter. In conclusion, piglets seem to prefer to sleep in a dark area during night time. Thus to attract piglets to the creep area it is better to keep it dark. The new radiant Eheater can heat up the creep area without light in contrast to the incandescent control lamp.

Development of a new farrowing pen for individually loose-housed sows: preliminary results for 'The UMB farrowing pen'

Inger Lise Andersen[1], Knut Egil Bøe[1], Donald M. Broom[2] and Greg M. Cronin[3]
[1]Department of Animal and Aquacultural Sciences, P.O. Box 5003, 1432 Ås, Norway, [2]Centre for Animal Welfare and Anthrozoology, Department of Veterinary Medicine, Madingley Road, Cambridge CB3 0ES United Kingdom, [3]Faculty of Veterinary Science, The university of Sydney, Rm No N105, Narellan 2567, NSW, Australia; inger-lise.andersen@nmbu.no

The objective of the present work was to collect preliminary production data on a newly-developed farrowing pen for individually loose-housed sows and to use these results to produce a pen for commercial practice which has a high piglet survival rate and improved sow and piglet welfare. The 'UMB farrowing pen' (7.9 m^2) comprises two compartments: a 'nest area' and an activity/dunging area with a threshold in between. The nest area has solid side walls, sloped walls on three of the sides and a hay rack on the fourth wall allowing free access to hay or straw. The nest area had two zones with floor heating covered by a 30 mm thick rubber mattress. Forty clinically health sows (28 Australian sows and 12 Norwegian, balanced for parity), were used in the experiment. The pens in the two different countries were the same except that rubber coating was not used in the activity area in the Australian pens during summer, The pens consisted of a nest area that was separated from a dunging area with a threshold between them and hay and straw for nest-building were supplied from a hay rack. Floor heating and sloped walls were used to stimulate sows and piglets to rest in preferred locations in order to reduce the likelihood of crushing. The preliminary results show that the production results were similar to, or better than, those reported for other types of pens for individually loose-housed sows. Mortality of live born piglets was 12.1±2.9% in the Norwegian sows and 12.9±2.0% in the Australian sows whereas the number of weaned piglets in both countries was 12.1±0.4 and 9.1±0.3, respectively. Overlying and starvation were the most common causes of death and were significantly affected by parity and litter size. All sows showed a high level of communication with their piglets, and primiparous sows communicated significantly more with their newborns during the birth process than the pluriparous sows. At parturition, 33.3% of the sows were resting with the back towards the back wall whereas 41.7% rested towards the threshold. In 50% of the nursings, the sows were resting against the back wall while 30% of the nursings occurred towards the threshold. This pen produced good production results under experimental conditions. Some changes in the threshold design and the depth of the nest area was needed for the commercial version of the pen.

Welfare implications of alternative housing systems for farmed meat rabbits

Rebecca Sommerville, Amélie Legrand and Vicky Bond

Compassion in World Farming, Food Business, River court, GU7 1EZ, United Kingdom; rebecca.sommerville@ciwf.org.uk

Rabbits are the second-most farmed species in Europe and fourth in the world, with more than one billion reared globally every year. In spite of this, no species-specific European Union legislation currently exists to protect them. Rabbits are most commonly housed in small barren cages, leading to serious welfare issues due to high stocking densities, disease and injury, wire flooring and a lack of behavioural opportunity in a barren environment. Development and uptake of alternative systems has been slower than in other farmed species. We undertook a review of existing alternative systems for rabbits in commercial practice across Europe. The aim was to identify which system or key features of a system would provide for all the needs of rabbits. Here we describe examples of existing commercial alternatives and discuss the implications for rabbit welfare. Conventional cages have been modified to include partial plastic flooring and gnawing blocks, but space and potential behavioural expression remain limited. In response to welfare concerns and consumer pressure, new indoor pens have been designed recently, known as the 'Park' or 'Barn' system. These provide a higher space allowance, plastic flooring, group housing and more environmental stimulation with platforms, tubes, gnawing blocks and a hay or straw dispenser. Research shows welfare benefits of these elements through increased feeding, exploration, activity and social behaviour, and reduced stereotypic behaviour. Bedding provision remains challenging due to the species-specific issue of increased mortality from enteric disease. Free range production exists, but remains limited to a minority of small scale, typically organic farms. Outdoor access is provided through mobile runs or paddocks with an indoor shelter. However, mortality rates can be high due to disease and predation. Alternative housing systems for breeding does are also being developed; but inter-individual aggression between does represents a major hurdle to the wide uptake of group housing. Changes in national legislation, industry and research developments, NGO pressure and adoption of alternative systems by food companies signal the start of 'moving on' from caged housing for farmed rabbits, though more research is needed into optimal housing conditions.

Relationship between keel mineralization and deformities of seventy-one week old White Leghorn hens

Patricia Y. Hester and Stacey A. Enneking
Purdue University, Animal Sciences, 125 South Russell St, West Lafayette, IN, 47907-2042, USA; phester@purdue.edu

Keel deformities and fractures in laying hens are welfare issues as previous research has reported that keel fractures cause pain. The objective of this study was to determine if a relationship exists between keel bone mineralization with deformities and fractures of the keel. Three-hundred and four keel bones were retrieved from 71 wk-old White Leghorns after hens were euthanized and individual BW were obtained. The keel was cleaned of muscle. Each keel was assessed for deviations by assigning a score ranging from 1 to 4. A score of 1 signified a severe deformity and a score of 4 represented a normal keel. Each keel was examined for fractures. The bone mineral density and bone area of each keel were determined using dual energy x-ray absorptiometry (DEXA). Correlation analysis was performed on bone traits and BW. The incidence of keel fractures was 88% with the majority being old fractures. The proportions of hens with keel deformity scores of 1, 2, 3, and 4 were 10.5, 12.2, 32.9, and 44.4%, respectively. Bone mineral density of the keel bone was correlated with keel deformity score ($r=0.25$, $P<0.001$), but not to keel fracture ($r=-0.08$). A negative correlation existed between keel deformity and the surface area of the keel bone that was scanned by DEXA ($r=-0.19$, $P<0.01$). Body weight was not correlated to keel deformity or fracture incidence. These results indicate that hens with severe keel deformities (score=1) have poorer bone mineral density (0.104 g/sq cm) and larger keels (bone area of 14.2 sq cm) than hens with normal keels (score=4, bone mineral density of 0.114 g/sq cm, and keel area of 12.71 sq cm). Implications are that hens with larger keels have less dense bones that can lead to severe keel deformities.

Can a bespoke management package reduce the risk of injurious pecking occurring in flocks of intact-beaked laying hens?

Jon Walton, Paula Baker, Sarah Lambton, Claire Weeks and Christine Nicol
University of Bristol, Animal Welfare and Behaviour, School of Veterinary Sciences, University of Bristol, Langford House, Langford, Bristol, BS40 5DU, United Kingdom; jon.walton@bristol.ac.uk

Injurious pecking (IP) in laying hens remains an important welfare and economic problem. It is common practice in most countries to beak trim the birds to reduce the impact of IP. However, this practice has been associated with both acute and chronic pain and is considered a mutilation in the EU. Beak trimming is also not a panacea for IP, since the behaviour is still evident in beak-trimmed flocks. Consequently, alongside other European countries, the UK is considering a ban on beak trimming. To inform this decision 20 commercial flocks of intact-beak birds were recruited for a study designed to establish whether such flocks could be kept with good welfare outcomes. Flock sizes ranged from 1200-16000 birds, and the majority of farmers had not previously kept intact-beaked birds before. By using and developing strategies found to effective in reducing the risk of IP in beak-trimmed commercial flocks, bespoke management plans were provided to both rearing and laying farmers. This included, but not limited to: strategies which improve litter quality and range use and enrichments to stimulate exploratory and foraging pecking inside and outside the house. Flocks were visited during the rearing period, at 8 weeks, and again during the laying period at 20, 40 and 65 weeks. Data collected include behavioural observations and plumage assessments, and production and mortality data for the study flock and for as many preceding flocks in the same house for which data were available. Acceptable limits for plumage damage and mortality were assigned prior to the start of the study. We will present early data on the ability of farmers using these management strategies to manage intact-beaked flocks.

Hair cortisol as an indicator of physiologically compromised status in dorcas gazelles

Oriol Tallo-Parra[1], Annais Carbajal[1], Maria Sabes-Alsina[1], Vanessa Almagro[2], Hugo Fernandez-Bellon[2], Conrad Enseñat[2], Miguel Angel Quevedo[3], Xavier Manteca[1], Teresa Abaigar[4] and Manel Lopez-Bejar[1]

[1]Universitat Autònoma de Barcelona, Veterinary Faculty, Edifici V, Campus UAB, 08193 Bellaterra, Spain, [2]Parc Zoològic de Barcelona, Parc de la Ciutadella, s/n., 08003 Barcelona, Spain, [3]Zoobotánico Jerez, Taxdirt, s/n, 11404 Jerez de la Frontera, Cádiz, Spain, [4]Consejo Superior de Investigaciones Científicas, Estación Experimental de Zonas Áridas, Crta. Sacramento, s/n, 04120 La Cañada de S. Urbano, Almería, Spain; oriol.tallo@uab.cat

The dorcas gazelle (*Gazella dorcas*) is an endangered mammalian classified as vulnerable for which captive-breeding programs and reintroductions are part of the actions addressed for their conservation. Increased cortisol secretion has been widely associated with stress and other physiological situations that may decrease animal welfare. Cortisol detection in hair is a new methodology capable to represent medium or long term circulating cortisol levels. In this study, hair cortisol concentrations from twenty-one dorcas gazelles (17 males and 4 females) from three different locations, Barcelona Zoo (n=6) and Zoobotanico Jerez (n=4) in Spain and the Gueumbeul Natural Reserve (n=11) in Senegal, were used. The average age was 2.97 years old (0.5 to 8.7 y.o.) and animals were classified as healthy (n=17) or physiologically compromised (n=4). Diagnosed disorders included two abortions, one necrobacillosis and one animal with several masses in the cervical region. Animals were also classified as leader (n=2) or not leader (n=19) after behavioural observations. All hair samples were harvested from the rump region and stored at room temperature until steroid extraction. After washing and mincing, samples were incubated with methanol for 18 hours at 30 °C and extracts were analysed using enzyme immunoassay kits (Neogen Europe, Ayr, UK). Assay validation showed a recovery percentage of 99.2% ±16.9, an intra-assay variability of 8.95% and R^2 from parallelism test of 99%. Sex, age, group and leader status did not have a significant effect over hair cortisol concentration, although females and leaders shown higher levels. The origin and health status of the individuals did significantly affect the cortisol levels (ANOVA test, P-value=0.010 and 0.010, respectively), with sick animals presenting the highest concentration (3.77±0.70 vs 2.40±0.72 pg/mg hair). Hair cortisol seems to provide helpful retrospective information about the physiological status of the animals in addition to housing conditions. Therefore, hair cortisol detection can be a potentially useful tool to employ in wild animal welfare assessment as an indicator of a physiologically compromised status.

Potential benefits of dog appeasing pheromone (DAP) to reduce stress in captive wolves (*Canis lupus*) during transport

Conrad Enseñat[1], Alessandro Cozzi[2], Patrick Pageat[2] and Xavier Manteca[3]
[1]Parc Zoològic de Barcelona, Parc de la Ciutadella s/n, 08003 Barcelona, Spain, [2]IRSEA, Le Rieu Neuf, 84, 490 Saint-Saturnin-lès-Apt, France, [3]Universitat Autònoma de Barcelona, School of Veterinary Science, Edifici V, Campus Universitari, 08193 Bellaterra (Barcelona), Spain; censenat@bsmsa.cat

Dog appeasing pheromone (DAP) is produced by suckling female dogs (*Canis lupus familiaris*) and previous studies have shown that it reduces fear and stress in both puppies and adult dogs. Wolves and dogs are remarkably similar in their genetic make-up and behaviour. The purpose of this paper is to report on the use of DAP to reduce the stress caused by transport of captive wolves. The study subjects were a group of five female captive-born Iberian wolves (*Canis lupus signatus*) made up by the mother and four daughters from two different litters. All five wolves were healthy and in good physical condition. The wolves were transported by road over 530 Km. On the day of transport the five wolves were captured and placed in five individual dog travel kennels that had not been previously used. The kennels were adapted to be used for wolves and thoroughly cleaned. The inside wall of the travel kennels were sprayed with DAP immediately before the wolves were placed into them. On arrival, wolves were kept in the transport kennels overnight and released the next morning. During the night, transport kennels were placed in a quiet area and covered by a blanket that had been sprayed with DAP. Before releasing the wolves in their new enclosure, the indoor area was thoroughly cleaned with water, a bicarbonate solution and an enzymatic detergent. The outdoor planted area was cleaned with a bicarbonate solution and both the indoor and the outdoor areas were sprayed with DAP. A DAP electric diffuser was placed in the indoor area. Wolves were monitored at regular intervals during transport and after their arrival. No evidence of panting, agitation or sudden movement was ever observed. The behaviour of the five wolves after being released in their new enclosure did not seem to be qualitatively different from what has been observed on other occasions when DAP was not used. However, the period of nervousness and agitation shown by the wolves upon release seemed to be much shorter than previously observed in other Iberian wolves transported without DAP. Taken together, our observations seem to suggest that DAP may be useful to reduce stress in captive wolves subjected to transport.

Clinical and ethological considerations in finch illegal capture rehabilitation

Maria Pifarré[1] and Cecilia Pedernera[2]
[1]Aiguamolls de l'Empordà WRC, IAEDEN, El Cortalet-PNAE, 17486 Castelló d'Emp., Spain, [2]IRTA, Animal Welfare subprogram, Fca Camps i Armet, 17121 Monells, Spain; pifam@hotmail.com

The capture of birds of Fringillidae family, in order to maintain them in captivity and train them to sing, is a traditional practise in Spain. This is a controversial practise because these birds' welfare is compromised from the capture process until their captivity living conditions. Although this is a legislated activity, most of the songbirds are taken from nature illegally. One of the most frequent methods used to capture them are glue traps, which leave the finches with fractures, altered conciousness, lack of feathers and glued one to another, or to branches or to themselves. From 2007 to 2013, 469 finches were treated at Aiguamolls de l'Empordà Wildlife Rehabilitation Center (WRC), 82% of them came from illegal captures. Frequently found species included: Carduelis cannabina(n=19), Carduelis carduelis(n=327), *Carduelis chloris* (n=48), *Fringilla coelebs* (n=2) and *Serinus serinus* (n=10). The clinical and behavioural evaluation of the birds arriving at the WRC consisted of the assessment of: attitude (relaxed, vocalizing, fluffed, ruffled or sleepy), reactivity, movement, signs of desequilibrium and grip on the perch, as well as weight, palpation and visual inspection. The regular admission protocol to treat these animals includes cleaning them from glue and reducing the state of stress. On August 2013 a group of nine *C. carduelis* and one *C. chloris* highly glued arrived at our WRC. *C. carduelis* mean $\pm$ SD ingress weight was 15.1$\pm$1.8 g and 24 g for *C. chloris*. Most of the birds showed signs of dyspnoea, were fluffed and sleepy, reduced reactivity, and didn't grip on the perch. After 24 hours of treatment three animals died and the rest showed normal behaviour (grip on the perch and eating) but 48 hours after their arrival five more birds were found dead. The necropsy showed that four presented intracranial haemorrhage and one skull fracture. Post-mortem mean $\pm$ SD weight was 10.1$\pm$0.9 g and 17 g for *C. chloris*. After 72 hours the two remaining animals were released. Due to the stress state of the captured birds arriving at the WRC, clinical assessment is based mainly on behavioural responses. Even when finches showed and improvement in their behaviour suggesting a recovery, most of them died after 48 hours. This apparent recovery could be only a strategy of the birds to show they are healthy. This behaviour may hide the real clinical state of the animal produced by a head trauma. Considering that head trauma can be another important and frequent consequence of finch captures as shown in this case report, an admission protocol including anti-inflammatory treatment with vitamin B complex and oxygen therapy support should be recommended.

Comparative eating behaviour of captive Asian elephants

Marc Pierard[1], Freek Van Riet[1], Freia Vercruysse[1], Lotte Vandyck[1], Evelien Driesen[1], Zjef Pereboom[2] and Rony Geers[1]
[1]*KU Leuven, faculty of Bioscience Engineering, Kasteelberg Arenbergpark 20, 3001 Heverlee, Belgium,* [2]*Royal Zoological Society of Antwerp, Centre for Research and Conservation, Koningin Astridplein 20-26, 2018 Antwerpen, Belgium; marc.pierard@biw.kuleuven.be*

Asian elephants (*Elephas maximus*) in captivity are managed in very different ways. We compared the eating behaviour of working elephants in Nepal (n=11) and a group at the animal park Planckendael in Belgium (n=6). In Nepal they received branches of Ficus, grass and kuchis (rice, salt and molasses rolled in grass). They were taken out into the national park regularly for grazing but this was not included in this study. In Planckendael the food was varied, consisting of hay, branches, fruits and vegetables. For statistical analysis food was divided into roughage and other feed. Elephants in Nepal were chained when not working or grazing. Those in Planckendael moved between an indoor and an outdoor enclosure. Videos of the elephants during main feeding moments were analysed with JWatcher. Time for chewing roughage and other feed and time manipulating roughage and other feed were measured and compared with Wilcoxon ranked sum test. Elephants in Nepal manipulated their food longer, both roughage (Nepal mean 44 s SD 37, zoo mean 13 s SD 7, P=0.014) and other feed (Nepal mean 39 s SD 34, zoo mean 11 s SD 8, P=0.036). For other feed the animals in Nepal also showed a trend to manipulate their food for a higher fraction of time while chewing (Nepal 0.34 SD 0.18, zoo 0.13 SD 0.089, P=0.06). There were no significant differences according to sex. No significant correlations were found but the sample size was small. Despite large differences in management and food composition, the eating behaviour of both groups showed similarities. More research is needed to look at morphological limitations as an explanation. The reasons for elephants in Nepal to manipulate their food longer are not clear. It could be related to housing and management or to differences in the food. More detailed studies are necessary.

How breeding and rearing practices affect welfare in the common marmoset (*Callithrix jacchus*)

Hayley Ash and Hannah M. Buchanan-Smith
University of Stirling, Psychology, Stirling, FK9 4LA, United Kingdom; hayley.ash@stir.ac.uk

The common marmoset is widely used in many areas of biomedical research, with many also housed world-wide for breeding purposes. As we have an ethical obligation to animals in our care, it is vital that their welfare is maximised, and they are 'fit for purpose' as research models. However, there are problems associated with breeding in captivity. Back-record data from three UK laboratory marmoset breeding colonies over four decades were examined, to investigate factors affecting dam longevity and litter size. Differences were found among sites, as well as over decades. Cox proportional hazards regression analyses revealed significant effects ($P<0.05$) of mean litter size and yearly production on dam longevity. There was however no consistent improvement in dam longevity over time, with the most frequent cause of death recorded being euthanasia due to 'poor condition'. Linear regression models revealed that no reproductive variable was useful in explaining mean litter size, except dam weight at conception (21.7%). While triplet litters were common, infants weigh less than smaller litters and there is high infant mortality, despite human intervention to improve survival. Such intervention also involves early separation from the family, which is known to have adverse consequences. Litter size at birth and rearing background were studied in relation to welfare. Infant behavioural development was recorded from birth to 8 weeks of age. Kruskal-Wallis tests revealed few major behavioural differences between litter sizes. Triplets, where only 2 remain with the family due to the loss of the third (2 stays), spent a significantly ($P<0.05$) higher percent of time in the suckling position (median 16%) than twins (median 12%), while singletons engaged in significantly more solitary play (median 0.9%) than 2 stays (median 0.3%). Adult behaviour in tests of temperament and affective state were then examined. While increased fear may be expected in individuals exposed to early life stress, preliminary ANOVAs suggest no difference in latency to obtain food from a novel object or a human between family-reared (mean 66 secs; 4.4 s) and supplementary-fed marmosets (mean 71 secs; 4.4 s). These findings will be compared to responses in cognitive bias tests. Evidence-based recommendations are discussed to improve colony management and enhance the welfare of animals in breeding facilities.

Experience of a landscape immersion exhibit: zoo visitors' perceptions of captive Japanese macaques (*Macaca fuscata*)

Azusa Yatsushiro[1], Misaki Furuie[2], Yasutaka Motomura[2], Atsushi Matsumoto[3], Masayuki Tanaka[4] and Shuichi Ito[2]

[1]Tokai University, Graduate school of Agriculture, Kawayo Minamiaso-mura, Aso-gun,Kumamoto-ken, 869-1404, Japan, [2]Tokai University, School of Agriculture, Kawayo Minamiaso-mura, Aso-gun,Kumamoto-ken, 869-1404, Japan, [3]Kumamoto City Zoological and Botanical Gardens, 5-14-2,Kengun,Higashi-ku,Kumamoto-shi,Kumamoto-ken, 869-0911, Japan, [4]kyoto City Zoo, Okazaki Koen,Okazaki Houshojicho,Sakyo-ku,Kyoto-shi,Kyoto-fu, 606-8333, Japan; 9anz2105@gmail.com

Recently, the zoo exhibit has been changing from a traditional wire cage style to a naturalistic style. Landscape immersion, a naturalistic style of animal exhibit, is intended to give visitors a sense that they are actually in the animals' habitat. At Kumamoto City Zoological and Botanical Gardens in Japan, a Japanese macaque exhibit was changed from an old cage-style exhibit to a landscape immersion exhibit in October 2013. It was thought that the change in the exhibit may affect the visitors' perception of the animals as well as the animal behaviour. We conducted a questionnaire survey to gather information about the perception of the Japanese macaques in both exhibition facilities. Zoo visitors who saw cage-style exhibits (n=566) and the landscape immersion exhibit (n=544) were asked to complete a questionnaire indicating their perception of (1) their excitement; (2) animal comfort; (3) the time spent watching; (4) animal intelligent and (5) animal visibility. Respondents were asked to select 5 of 10 species of primates including Japanese macaques, being reared in Kumamoto Zoo, and to rank them according to perception. For example, 'Please arrange species in an order that indicates the level of comfort of life for this species.' The respondents were not informed about the Macaque survey. In addition, we interviewed visitors who had seen the Japanese macaques living in the two different exhibits. The results of the questionnaire survey indicated the following: the change of exhibit from the cage style to the landscape immersion style resulted in visitors' impression that the comfort of life of the Japanese macaques was increased (chi-square test: $\chi^2=200.3$, df=5, P<0.001), time spent watching was increased ($\chi^2=68.02$, df=5, P<0.001) and the animal visibility in the exhibit was increased ($\chi^2=68.02$, df=5, P<0.001). However, visitors' perception of the amount of excitement they felt or the intelligent of the Japanese macaques did not change. Results of the interview survey were similar to the above results. Our survey suggests that the landscape immersion exhibit affected the perception of animal comfort, and the time visitors spent observing the Japanese macaques, but it had little effect on how exciting or intelligent they perceived the exhibit to be.

Calving behavior in Holstein dairy heifers and cows

Júlio O. J. Barcellos[1], João Batista G. Costa Jr.[1], Maria Eugênia A. Canozzi[1], Zach Weller[2], Richard K. Peel[2], Jack C. Whittier[3] and Jason K. Ahola[3]
[1]Universidade Federal do Rio Grande do Sul, Departamento de Zootecnia, Porto Alegre, RS, 15100, Brazil, [2]Colorado State University, College of Natural Science, Statistic, Fort Collins, CO, 80523, USA, [3]Colorado State University, Department of Animal Sciences, Fort Collins, CO, 80523, USA; julio.barcellos@ufrgs.br

The aims of the study were to determine relationship between calving process (CP), calf birth weight (BW), and calf sex in Holstein dairy heifers (n=89) and cows (n=105). It was approved from the Colorado State University Animal Care and Use Committee (protocol #11-2583A). Data were collected during CP for each female. This included time from start to lay down to visibility of water bag (CP–T1); from visibility of water bag to start expelling the calf (CP–T2); from start expelling the calf to appearance of the calf's feet (CP–T3); and from appearance of the calf's feet to actual delivery of calf (CP–T4). Information of calf included BW and sex. Time intervals from first signs to actual parturition were analyzed using PROC MIXED with time as a repeated measure within heifer. BW, calf sex, CPs, and interactions, and BW and CPs were fixed effects and compared by SCHEFFE'S TEST. Spearman correlations coefficients were calculated among CPs, and BW. There was a significant difference ($P<0.05$) for each step of CP for heifers and cows, and CP–T4 average time was 150.86 min for heifers and 111.7 min for cows. Heavier calf was positively correlated with average time to CP–T1 (0.33) for heifers. A positive correlation was found for heifers and cows between CP–T1 and CP–T2 (0.62 and 0.77), CP–T1 and CP–T3 (0.58 and 0.66), CP–T1 and CP–T4 (0.38 and 0.53), CP–T2 and CP–T3 (0.81 and 0.85), CP–T2 and CP–T4 (0.57 and 0.61), and CP–T3 and CP–T4 (0.72 and 0.78). Different strategies of calving management should be used for heifers and cows such as increased monitoring during CP and use lighter sires for heifers to prevent long CP which may result in difficult calving.

Using epidemiology to study behaviour: examples from guide dogs

Lucy Asher, Gary England and Martin Green

University of Nottingham, School of Veterinary Medicine and Science, Sutton Bonington, LE12 5RD Leicestershire, United Kingdom; lucy.asher@nottingham.ac.uk

Epidemiology is the study of patterns of health related states or events in populations. Statistical models developed for studying epidemiology can be applied to behavioural states or events. Models have been designed for discrete data, which is common in ethology, and to incorporate time. The application of epidemiological statistical models to behaviour will be presented here, specifically survival analysis and multi-state modelling applied to data on guide dogs. Survival analysis, as the name suggests, was developed to examine mortality data and risks associated with time until death. We applied a cox proportional hazards model to examine the time taken to withdraw a dog from guide dog training that was not behaviourally suited to the guiding role. We found sex, breed and other factors affected time to withdrawal. Bitches were withdrawn faster than dogs (by a median of 1.5 weeks, P<0.0001); Labradors were withdrawn faster (2 weeks, P=0.05), and Labrador cross Golden Retrievers slower (2.5 weeks, P=0.008), than Golden Retriever cross Labradors; and dogs not bred by Guide Dogs were withdrawn faster than those bred by Guide Dogs (3.1 weeks, P=0.003). Survival analysis can be applied to a wide range of behavioural data including: latency to choose in a choice test; time to rehoming; time to learn an association; age of development of a behaviour. Extending survival analysis, multistate modelling can be used to incorporate more than two events or states. Multistate modelling has classically been used to understand transitions between disease states. When studying cancer, epidemiologists have considered the rate at which patients move between states of treatment for cancer, remission and death. We used multistate models to investigate transitions between states of training to qualification as a guide dog or behavioural withdrawal, and from qualification as a guide dog to behavioural withdrawal. We found that the sex, Breed (with Labradors=Golden retriever≠ F1 crosses of Labradors and Golden Retrievers) and whether the dogs were bred by Guide Dogs effected movements between these states (all P<0.0001). Multistate modelling has already been applied to understand transition between mobility scores in livestock, but could be applied to understand a wide range of short- and long-term changes in states. For example, to understand transitions between lying, standing and moving in relation to lameness or production; or to understand animal decisions, where the current choice may depend on previous choices. There are many more possibilities for applying such methods to behavioural data and it is hoped that this talk will encourage more people to utilise these, either themselves or through collaboration with statisticians or epidemiologists.

Environmental-induced deficit of memory in female gerbils

Katarzyna Zieba
Uniwersity of Warsaw, Department of Animal Behaviour at Faculty of Psychology, Warsaw, Poland; kzieba@psych.uw.edu.pl

Everyone who wants to work with animals should be familiar with the general principles of breeding and experiments, and should know how to prepare for a research taking advantage of previously conducted tests. Despite the fact that animals are provided with welfare, there is still a problem of conditions in which the animals live. Legal regulations specify basic conditions, which should be fulfilled by a research or breeding unit as far as care and maintenance is concerned. However, the regulations do not standardize all factors. This attitude can influence test results, which are conducted by different researchers. Many researchers minimize description of conditions of keeping the animals as much as they can. Dimensions of the cage are given rarely, let alone the number of animals inside. How frequent were the contacts between a researcher and an animal; was there anything except sawdust in the cage? Presented research consisted of 40 gerbils of both sexes which were kept in 5 different groups: I. control, which was a group of four animals in cage of type IV (Tecniplast). The cage was filled only with sawdust, and there was a special area on the top where animals could use wholesome and balanced ad libitum food. I treat the control group as a standard for keeping animals in a laboratory. II. enriched environment group, which besides the conditions of the control group offers wooden blocks for blunting incisors and for fun as well as hydraulic elbows for hiding. III social deprivation – only one animal was placed in the cage at a time. IV handling – the animals were carried and stroked for about ~7 minutes every day. V. nutritional deprivation – is a group, in which the amount of food was reduced and weight of the animals was maintained at 10% less than regular. After a month the animals underwent one of the behavioural test to measured memory. Barnes maze is a circular table with 20 circular holes – only one of them was an 'escape box'. Each of the animals was placed into the maze in 4 consecutive days and next, in 8th day. In 4th and 8th day of experiment the 'escape box' was removed. In day 8, we measured path length to the place where the 'escape box' was originally placed. There was a statistically significant interaction effect of both factors: sex and group, $F_{4.30}=4.55$; $P<0.001$; h=0,38. Bonferroni correction shows that inside handling and food deprivation groups were different, respectively: $F_{1.30}=4.68$; $P<0.05$ and $F_{1.30}=7.86$; $P<0.01$. In both groups females achieved the worst results than males. Females in enrichment group had significantly better results than in food deprived group $F_{4.30}=3.26$; $P<0.05$. There were no statistically relevant differences in male group. Research financed by National Finansed Centre DEC-201101NHS600673.

Regional brain neurotransmitter levels: sex, halothane genotype and cognitive bias

Laura Arroyo[1], Daniel Valent[1], Raquel Peña[1], Anna Marco-Ramell[1], Mireia Argany[1], Antonio Velarde[2], Josefa Sabrià[3] and Anna Bassols[1]
[1]*Universitat Autònoma de Barcelona, Bioquimica i Biologia Molecular. Facultat de Veterinària, Edifici V. Campus UAB, 08193 Cerdanyola del Vallès, Barcelona, Spain,* [2]*IRTA, Animal Welfare, Veïnat de Sies s/n, 17121 Monells, Girona, Spain,* [3]*Universitat Autònoma de Barcelona, Bioquimica i Biologia Molecular and Institut de Neurociències, Edifici M. Campus UAB, 08193 Cerdanyola del Vallès, Barcelona, Spain; anna.bassols@uab.cat*

The development of methods for assessing the affective (or emotional) states is a crucial step in improving animal welfare. The 'cognitive bias', defined as a pattern of deviation in judgment in particular situations, is used as a label (optimistic or pessimistic) for the effects of affective state on cognitive processes. The aim of this study was to determine the concentration of indoleamines (5-hydroxyindole-3-acetic acid (5-HIAA) and serotonin (5-HT)) and catecholamines (noradrenaline (NA), dopamine (DA), 3,4-dihydroxyphenylacetic acid (DOPAC) and homovanillic acid (HVA)) in the amygdala, hippocampus and prefrontal cortex of a group of slaughtered pigs classified according their emotional state, sex and halothane genotype. The study was carried out on 48 hybrids Large White × Landrace pigs housed at Institut de Recerca i Tecnologia Agroalimentàries (IRTA)-Monells facilities. Animals were trained to learn to discriminate positive and negative spatial cues and classified according to their emotional state during the cognitive bias test. Regional distribution of brain monoamines showed similar patterns to those described in the literature and our previous results. Between sexes, halothane-carrier males showed lower concentration of dopamine and its metabolites in the amygdala, whereas halothane-carrier females showed higher concentrations of serotonin and its metabolites in the hippocampus. When considering the effect of cognitive bias on neurotransmitter profile, females defined in a positive emotional state showed higher concentrations of homovanillic acid (P<0.05), a metabolite of dopamine, in the amygdala, probably related to the tendency to decrease in dopamine concentration found in these animals. This suggests a relationship between motivation and cognition. We conclude that halothane genotype produces changes in neurotransmitter's concentration of slaughtered pigs. Furthermore, we also found differences in dopamine metabolism in the amygdala between female pigs defined as optimistic or pessimistic.

Temperament traits in sheep predicted by strength of lateralisation

Lindsay R Matthews[1,2], Shanis Barnard[1], Stefano Messori[1], Luca Candeloro[1], Michele Podaliri Vulpiani[1] and Nicola Ferri[1]

[1]*Istituto Zooprofilattico Sperimentale dell'Abruzzo e del Molise 'G. Caporale', Human-Animal Relationship and Animal Welfare Laboratory, via Campo Boario, 64100 Teramo, Italy,* [2]*Lindsay Matthews & Assoc Research International, via Colle Morino 8, Scerne di Pineto, 64025 Teramo, Italy; lindsay.matthews1@gmail.com*

Left lateral biases have been associated with increased emotional responsiveness in several species, reflecting right hemispheric (brain) control of intense, negative events. Alternatively, in other studies, emotional reactivity has been associated with strongly lateralised behaviour, regardless of the directionality of the bias. Our aim was to determine if left laterality or the absolute magnitude of laterality in sheep would be the best predictor of reactivity during a simulated common farming practice (i.e. isolation and forced separation of ewes and their lamb during artificial weaning). Laterality was assessed in individual-animal tests (43 ewes or their singleton lambs) by recording the direction of deviation upon encountering a centrally-located obstacle in a laneway when returning flockmates. Behavioural reactivity of lambs and dams was assessed over 5 minutes when they were physically separated by an open-mesh fence in adjacent pens. Time spent near the fence (within 80 cm), vocalisations and number of times the animal moved toward and away from the fence were recorded. Nineteen ewes (44.2%) and 35 lambs (81.4%) showed individual laterality (binomial test, $P<0.05$). A Mann-Whitney U test showed that lateralised lambs spent significantly more time near the fence (247.6 ± 49.3 s) than non-lateralised animals (189.9 ± 63.0 s) ($U=216$, $P=0.02$), but were not more active or vocal. Arguably, the lateralised animals showed greater attraction to the dam because they are more disturbed and, thus, required greater reassurance from her. The data show that the absolute magnitude of laterality is a potential novel predictor of separation stress and/or the degree of alleviation provided by the mother.

Litter loss in laboratory mice – do mothers kill their pups?

Elin M Weber[1,2], Bo Algers[2], Jan Hultgren[2] and I Anna S Olsson[1]

[1]*IBMC-Instituto de Biologia Molecular e Celular, Universidade do Porto, Laboratory Animal Science, Rua Campo Alegre 823, 4150-180 Porto, Portugal,* [2]*Swedish University of Agricultural Sciences, Department of Animal Environment and Health, P.O. Box 234, 532 23 Skara, Sweden; elin.weber@slu.se*

Successful mouse breeding is a crucial part of providing animals for research. Perinatal mortality is a relatively common problem; reported pup mortality risks until weaning vary from nearly 0 to 50% in experimental studies. However, little is known about the way pups die. Large numbers of animals dying constitutes a welfare problem and leads to an increase in the number of breeding animals needed. Dead pups are generally eaten, and it is commonly assumed that the mother has killed her pups. The aim of this study was to increase the understanding of pup mortality and investigate if females actively kill their pups. We used video recordings of 10 primiparous female mice (strain C57BL/6 and knockouts Hfe$^{-/-}$ and β2m$^{-/-}$) that lost their litters, to observe the interactions between mother and pups in detail from parturition until the pups died. An ethogram was used for scoring behaviours, and a flowchart was used to systematically narrow in on certain time periods. Six of the females had pups that were never seen moving, indicating that some pups were most likely still-born. Females interacted with both moving and still pups, but were never observed manipulating a moving pup that stopped moving after the interaction, nor were any wounds visually detected on the pups during the video observations. Hence, in spite of very detailed observations, we could not verify that females kill their pups. These findings highlight the importance of being cautious when concluding how laboratory mouse pups die, and stress the need for more systematic investigations. By assuming that pups are killed by the mother, based on the fact that they are found half-eaten or not found at all, the true causes of pup loss are probably overlooked, and a welfare problem in laboratory mice is left unresolved.

Circadian rhythms control red jungle fowl crowing

Shuichi Ito[1], Makiko Hirose[1], Azusa Yatsushiro[2], Atsushi Matsumoto[3], Ken-ichi Yayou[4], Masayuki Tanaka[5] and Tsuyoshi Shimmura[6]
[1]Tokai University, School of Agriculture, Kawayo, Minamiaso-mura, Aso-gun, Kumamoto-ken, 869-1404, Japan, [2]Tokai University, Graduate School of Agriculture, Kawayo, Minamiaso-mura, Aso-gun, Kumamoto-ken, 869-1404, Japan, [3]Kumamoto City Zoological and Botanical Gardens, 5-14-2, Kengun, Higashi-ku, Kumamoto-shi, Kumamoto-ken, 862-0911, Japan, [4]National Institute of Agrobiological Sciences, Division of Animal Sciences, 2, Ikenodai, Tsukuba-shi, 305-8602, Japan, [5]Kyoto City Zoo, Okazaki Koen, Okazaki Houshojicho, Sakyo-ku, Kyoto-shi, 606-8333, Japan, [6]National Institute for Basic Biology, 38 Nishigonaka, Myodaiji, Okazaki-shi, Aichi-ken, 444-8585, Japan; shuichi_ito@agri.u-tokai.ac.jp

The crowing of roosters is frequently observed in the early morning. An internal biological clock, which works to a rhythm of 23.7 h, regulates the crowing that begins before dawn. The red jungle fowl, ancestor of the chicken, begins to crow at predawn in the wild. However, because it is not possible to assess the effect of ambient light in the wild, the mechanism that regulates crowing in the red jungle fowl has not yet been identified. In this study, we tested whether the crowing of red jungle fowl is controlled by an internal biological clock (i.e. circadian clock) or by external light stimuli. We exposed three red jungle fowls (No.1, No.2 and No. 3) in individual cages to 12 h of bright light and 12 h of dim light (12L12dimL) for 10 days. During the next 10 days, the birds were kept exposed to constant dim light (dimLL). The bright light condition was adjusted to 100 lux, whereas the dim light condition was maintained at 1 lux. Crowing of each fowl was recorded all day using a digital video camera equipped with a near-infrared illuminator. The movie was analysed for the incidence and duration of crowing per individual. In the 12L12dimL cycle, crowing was observed approximately 2.2 h before the onset of light (i.e. anticipatory predawn crowing), consistent with the observations of red jungle fowl in the wild. For 10 days, the birds crowed on an average of 3.05 h (No. 1), 2.14 h (No. 2) and 1.51 h (No.3), before the onset of light. Thereafter, under the dimLL conditions, a free-running rhythm of crowing was observed. The free-running crowing continued for 23.55 h (No.1), 23.8 h (No.2) and 23.8 h (No.3), which equated to 23.7±0.1 h (n=3) when subjected to periodogram analysis. This free-running rhythmicity gradually damped out. Our observations suggest that the red jungle fowl breaks the dawn every morning as a function of his circadian clock.

Animal based measures and transport conditions during long journey transport of calves

Cecilia Pedernera[1], Antonio Velarde[1], Beatrice Mounaix[2], François Raflegeau[2] and Hans Spoolder[3]
[1]IRTA, Animal Welfare Subprogram, Monells, 17121, Spain, [2]Institut de l'Elevage, Service Santé et Bien-être des Ruminants, 35652 Le Rheu Cedex, France, [3]WUR, Livestock Research, P.O. Box 65, Lelystad, 8200 AB, the Netherlands; cecilia.pedernera@irta.cat

In Europe weaned dairy calves are transported for long distances to be fattened in another country. Transportation of these animals can be terrestrial or a combination of road and ferry transport. The aim of the study was to assess calves welfare at arrival to control posts or assembly centres after long journeys using a protocol including animal based and resource and management based measures. The testing of this protocol was part of the project 'Development of EU wide animal transport certification system and renovation of control posts in the European Union', funded by DG SANCO. Seventeen trucks transporting a total of 4,811 small calves (50-70 kg) were assessed during unloading and in the pens. The average transport time prior to assessment was 16 hours (8-27) (min-max) and the average number of animals by truck was 283 (199-350) (min-max). As requested by the 1/2005 EU regulation, all truck ramps were covered with anti-slip flooring material, with foot battens. However, straw covering ramps was often not optimal. The ramp slope complied with the relevant regulation, except in 4 transports where it was steeper, and no sharp edges or damaged area or poor drained areas were observed on the ramp flooring. Gaps were observed between the ramp and the floor in 65% of transports. During unloading slipping was observed in 5.6% of the calves, and falling was only observed in 1.8%. Lighting during unloading was adequate for both the animal orientation and for the handlers, but blocking zones (mainly shadows and dead ends) were present in 66.7% transports. Deck height was considered adequate to allow for natural head posture, and bedding in the truck (mainly straw) was sufficient. Main risks for animal welfare were observed during handling, as stressful handling practices were observed in some cases. This included shouting and slapping animals, and also forbidden practices like pulling animals from the tail observed in 6 trucks. Three out of 4,811 calves were found dead on arrival or in the resting pen shortly after the unloading, and 2 of them were also scored as exhausted. In the pens no animals were observed with very low BCS. The proportion of shivering or panting was low <1% on average, as well as animals with diarrhoea <5%. Although the observed transport conditions were of overall good quality, with good resources management, calf welfare can be improved and continues to be an underestimated issue which deserves further research.

Identified exciting factors of soil on behaviour of fattening steers reared indoor

Sayuri Ariga[1], Shigehumi Tanaka[2], Takashi Chiba[2], Kyoichi Shibuya[2] and Shusuke Sato[1]
[1]Tohoku University, Graduate School of Agricultural Science, Laboratory of Land Ecology, 232-3, Yomogita, Naruko-Onsen, Osaki, Miyagi, 989-6711, Japan, [2]Tohoku University, Graduate School of Agricultural Science, Field Science Center, 232-3, Yomogita, Naruko-Onsen, Osaki, Miyagi, 989-6711, Japan; s_ariga220@yahoo.co.jp

For improving animal welfare, it is recommended to provide outdoor exercise area (OEA) for fattening cattle reared indoor. We previously demonstrated OEA of soil floor promoted normal behaviour and exercise than those in OEA of concrete floor and grassland in steers (2012). In this study, we provided soil in four types of substrate for steers, to identify exciting factors of soil on their behaviour. 12 steers were used (24 ± 2 months of age; body weight, 565 ± 60 kg; 3 steers/pen; 25 m^2/pen). Smell of soil (SME), pile of soil (PIL), tilled fluffy soil floor (TIL), and compressed soil floor (COM) were set in OEAs (25 m^2/OEA). We found there were no significant different effects between 1 and 5 hours access to OEA (2010). Therefore, in this study we provided each substrate only for 1 hour. Exploring behaviour (%/h) was observed by 5-min scan-sampling, and analysed by ANOVA. Display behaviour (No.), solitary play behaviour (No.) and social behaviour (No.) were observed by continuous sampling, and analysed by Kruskal-Wallis test. Steps per hour (No.) was recorded by pedometers, and analysed by Kruskal-Wallis test. We used data of 1 hour access to OEA of concrete floor in our study (2012) as Control. Exploring behaviour was more in PIL than Control ($62\pm18\%$ and $37\pm14\%$: $P<0.05$). Display behaviour was more in PIL and TIL than Control (23 ± 16, 2 ± 3, and 0 ± 0: $P<0.05$). Solitary playing, allogrooming, and steps were more in TIL and COM than Control ($P<0.05$). Steps were more in PIL, TIL, and COM than Control in the day after providing each substrate ($P<0.05$). Exploring behaviour was more in COM than other substrate ($P<0.05$). There was no effect by olfactory stimulation. We speculated it is possibility to promote more normal behaviour by combining with PIL and COM.

More solid feed vs. straw rack: best method to improve veal calf welfare

Laura Webb[1], Kees Van Reenen[1], Harma Berends[2], Bas Engel[3], Walter Gerrits[2], Imke De Boer[1] and Eddie Bokkers[1]
[1]Animal Production Systems Group, P.O. Box 338, 6700 AH Wageningen, the Netherlands, [2]Animal Nutrition Group, P.O. Box 338, 6700 AH Wageningen, the Netherlands, [3]Biometris, P.O. Box, 6700 AC Wageningen, the Netherlands; laura.webb@wur.nl

Veal calf diets provide insufficient rumination opportunities as indicated by abnormal oral behaviour development in these calves; i.e. tongue playing/rolling and excessive oral manipulation of environment. We investigated whether veal calves would benefit more from increased amounts of a solid feed mixture or unlimited availability of long straw. Holstein-Friesian bull calves (n=60) received milk replacer and solid feed (80% concentrate, 10% corn silage, and 10% chopped straw, DM basis) up to 27 wk of age. Control calves (CO) and calves with unlimited straw (ST) received 115 kg solid feed (based on Dutch veal). Calves with increased solid feed (SF) received 300 kg (based on intake of calves fed ad libitum). Milk replacer was adjusted for equal growth. ST calves received unlimited long barley straw in racks. Behaviour was monitored at 15 and 24 wk of age using instantaneous scan sampling every 2 hours from 06:30 to 20:30 h at a 6 min interval. Data were analysed with a generalised linear mixed model. ST calves used the straw rack 2.3±1.0 and 8.6±1.8% of total scans at 15 and 24 wk. ST calves displayed less abnormal oral behaviours than CO and SF calves (ST: 4.4±0.8, CO: 10.0±1.1, SF: 8.6±0.8; P=0.004). ST calves ruminated more than CO calves at 15 wk (ST: 13.0±0.9, CO: 8.6±1.0, SF: 10.3±1.4), and than CO and SF calves at 24 wk of age (ST: 16.5±2.0, CO: 6.0±0.7, SF: 8.3±1.1; P=0.043). No differences in tongue playing and lying levels were observed. Calves played more at 15 compared with 24 wk (15 wk: 1.8±0.3, 24 wk: 0.6±0.1; P=0.008), irrespective of treatment. This study suggests that unlimited straw has the potential to improve calf welfare more than increased amounts of solid feed, potentially because the straw offered unlimited chewing opportunity with no great impact on nutritional status.

Effect of positive interactions on parameters of the human-animal relationship in dairy calves

Stephanie Lürzel, Charlotte Münsch and Susanne Waiblinger
Institute of Animal Husbandry and Animal Welfare, University of Veterinary Medicine Vienna, Veterinärplatz 1, 1210 Vienna, Austria; stephanie.luerzel@vetmeduni.ac.at

The animal's relationship to humans is important for animal welfare and productivity. In order to test positive interactions (PI) as a means of improving the human-animal relationship in the short and medium term, we compared group-housed dairy calves that had experienced additional PI with calves that had not. The experiment was conducted at a commercial dairy farm with 77 Holstein-Friesian calves. At time of first testing, calves were housed in groups of 12-13 animals, later in groups of 25-26 animals. Standard care and management practices were applied to all calves; PI calves additionally experienced PI (stroking of ventral neck, gentle talking) for 3 min/d during the first 14 d of life. Tests (avoidance distance, AD – all calves; arena test – 23 PI, 21 control) were conducted after the treatment phase (mean age 18 d), and two further AD tests after disbudding (32 d) and weaning (86 d). The arena test comprised three phases of 3 min each: isolation, experimenter in the arena, isolation. The arena was subdivided in 5 areas (entrance, A, B, C, experimenter). After treatment, PI calves had lower ADs (median: 15,3 cm) than controls (40 cm; repeated measures ANOVA: $P=0.017$). After the painful procedure of disbudding, ADs were higher (medians of both groups: 70 cm; testing day: $P<0.001$, testing day × treatment: $P=0.002$). At 86 d of age, animals were tested by two observers. Mean AD tended to be lower in PI calves than in control if tested by observer 1 (ANOVA: treatment: $P=0.07$, observer: $P<0.001$, treatment × observer: $P=0.08$). During phase 2 of the arena test, PI calves spent more time in area C than control calves (Mann-Whitney U test: $P=0.001$), and control calves tended to spend more time in the entrance area than PI calves ($P=0.06$); thus, PI calves were closer to the experimenter. In sum, PI improved the human-animal relationship, indicated especially by the lower ADs after treatment. This effect may be persistent in the medium term even after an aversive experience like disbudding.

Effect of feeding frequency and total amount of milk replacer on young calves

Ellen Jongman[1] and Andrew Fisher[2]
[1]Animal Welfare Science Centre, Melbourne School of Land and Environment, University of Melbourne, Parkville 3010, Australia, [2]Faculty of Veterinary Science, University of Melbourne, Parkville 3010, Australia; ejongman@unimelb.edu.au

This study compared the effects on 46 peri-natal Friesian calves (±44 kg) receiving milk replacer at either: (1) 10% of body weight daily over one meal; (2) 10% of body weight daily over two meals; or (3) 20% of body weight fed over two meals. The milk was delivered in teat feeders which calves had access to for 1 hr/meal. Calves, housed in large individual pens on straw bedding, were studied over 5 replicates with 8-11 calves/replicate and each treatment represented at least twice/replicate. They arrived at the research facility in their second day of life and were studied for 7 days. Measurements included daily bodyweight, milk intake/meal, time spend drinking, non-nutrient suckling, time spend active, lying behaviour, behaviour during a novel arena test and blood sampling (at 07:00 h, 10:00 h, 13:00 h, 19:00 h, 22:00 h, 01:00 h and 07:00 h) on selected days for cortisol, PCV, glucose, insulin, NEFA and BOHB, indicative of stress, hydration and metabolic state. The three treatments were compared using REML analyses. The data shows that calves that were fed once daily compared to twice daily had lower feed intakes during day 3 and 4 of life ($P<0.001$). Calves fed at 20% of their body weight had higher milk intakes from the 4th day of life ($P<0.001$) and gained more weight compared to calves fed at 10% bodyweight (3.23 vs 0.65 kg; $P<0.001$). Non-nutrient suckling was mainly directed at the empty feeder immediately after feeding and showed significant differences both associated with the frequency of feeding and the amount fed (11 vs 22 bouts in calves fed at 10% over either one or two meals vs 10 bouts in calves fed at 20% over two meals on the 7th day of life; $P<0.01$). This may indicate that feeding at 10% did satisfy feeding motivation less than feeding at 20%, although no other behaviour indicators were affected by treatment. There were significant effects on physiological indicators of metabolic state, with generally higher concentrations of insulin and NEFA throughout the day ($P<0.05$) in calves fed at 20% and lower concentrations of glucose during night time in animals fed once daily ($P<0.05$). Cortisol was higher in calves fed once daily 3 hours after the other calves received their second feed (63 vs 42 and 38 mmol/l, $P<0.01$). It was concluded that feeding twice daily offers benefits to calves up to day 4 of life while feeding 20% of bodyweight was beneficial after day 4 to satisfy feeding motivation and nutrition for growth. While there were significant differences in metabolic measures with treatment, all measures were within normal physiological range, and the consequences for calves' welfare are not clear.

Dairy cow preference for pasture is affected by previous experience

Priya Motupalli, Emma Bleach, Liam Sinclair and Mark Rutter
Harper Adams University, Department of Animal Production, Welfare, and Veterinary Sciences,
Newport, Shropshire, TF10 8NB, United Kingdom; pmotupalli@harper-adams.ac.uk

Research has shown that previous experience of pasture affects the preference of non-lactating heifers for pasture, but the effect of previous experience on lactating cow preference is less well known. The aim of this study therefore was to determine if exposure to pasture during rearing influenced dairy cow preference for pasture. First-lactation, in-calf, Holstein-Friesian dairy cows (n=12×3) were reared in one of three groups: group A (zero-exposure to pasture during rearing, 2011-2013), group B (reared without exposure to pasture for their first grazing season in 2011 and with exposure to pasture for their second grazing season in 2012), group C (maximum exposure to pasture during rearing – exposure during first and second grazing seasons, 2011-2013). Cows were tested for preference in groups of three for 10 days using pair-wise comparisons over six repetitions during summer 2013. At the start of each replicate, cows were balanced by lactation stage. Comparison order was established using a Latin-square design. Cows had continuous access between one of two adjacent areas of housing (85.1 m^2, 1.3 free stalls/cow) and one of two adjacent 0.22 ha pastures via a 38 m track. Ad libitum TMR was available indoors i.e. as in commercial systems. Behaviour was recorded using scan sampling on days 1, 2, 4, 6, 8 and 10 of each measurement period during daylight hours. A one-way ANOVA and post –hoc Bonferroni tests revealed that A cows spent more time indoors than B or C cows (94.2 vs. 45.9 vs. 44.8%, P<0.001), B cows spent the least amount of time lying (41.4 vs. 33.4 vs. 40.8%, P=0.004) and ruminating (31.3 vs. 25.3 vs. 30.5%, P<0.001), and A cows spent less time grazing then either B or C cows (1.2 vs. 21.5 vs. 21.6%, P<0.001). In conclusion, dairy cows without exposure to pasture prefer being indoors over cows with exposure to pasture, and the time of exposure (year 1 vs. year 2) does not influence time spent at pasture or grazing times as both B and C cows spent the same amount of time at pasture and grazing.

The sickness response in bovine respiratory disease at peak clinical illness

Rachel L. Toaff-Rosenstein[1], Laurel J. Gershwin[1], Adroaldo J. Zanella[2] and Cassandra B. Tucker[1]
[1]University of California, 1 Shields Ave., Davis, CA, USA, [2]University of São Paulo, Pirassununga, SP, Brazil; rtoaff@ucdavis.edu

Bovine Respiratory Disease (BRD) is a top cause of morbidity and mortality, due in part to poor detection. The sickness response, elicited by inflammation, includes fever, anorexia, and reduction in activity not essential for short-term fitness, such as grooming. Response monitoring may improve BRD detection, especially if continuous. The objective was to quantify the magnitude of the sickness response at both peak illness and relative to disease severity. It was hypothesized that the response would reflect BRD, and more severe illness result in greater changes than mild sickness. Steers (7-10 m old) were given 1 of 5 BRD pathogens (BRD; n=20) or sterile media (Control; n=20) and housed by treatment (n=4/pen) with a grooming brush. Rectal temperature (TEMP, every 5 min, indwelling loggers), bunk attendance (BUNK, every 5 min, video), and self-licking (LICK, continuously, video) were monitored 24 h/d, while brush use (BRUSH, continuously, video) was recorded 13 h/d. A clinical sum score (CSS) was generated from daily exams of all steers. BRD steers were necropsied, recording portion of grossly affected lung (%LUNG). Data from 3 d were analyzed: peak illness (day of average highest CSS; PEAKDAY) and 2 d before. Generalized linear and mixed models were used to evaluate the effect of BRD (n=5 pens/treatment). BRD pens had higher TEMP (40.1 vs. 39.8±0.10 °C, P=0.04) and tendency for lower BUNK (1.5 vs. 2.1±0.18 hr/24 hr, P=0.06) compared to Control. Individual clinical health was variable (CSS range; BRD=151-665, Control=21-391); further analyses used steer-level data (accounting for group-housing) to describe the relationship between the sickness response and degree of illness, as reflected by CSS (n=40, except for LICK, n=30) and %LUNG (n=20; range=0.5-55%). Clinically sicker steers had a more marked response; those with higher CSS had greater TEMP (P<0.01; R2=0.64), lower BUNK (P<0.01; R2=0.35) and less LICK (P=0.02; R2=0.24) than those with lower CSS over all 3 d, accounting for a significant amount of variability on PEAKDAY. BRUSH was lower among the sickest steers on PEAKDAY only (CSS*day; P=0.04; R2=0.14). Similarly, BRD steers with greater %LUNG had an increased sickness response, including lower BUNK (P<0.01; R2=0.22), BRUSH (P=0.04; R2=0.22), and LICK (P=0.04; R2=0.26) than those with lower %LUNG over all 3 d, and accounting for a significant amount of variability on PEAKDAY. TEMP was also higher among those with greater %LUNG on PEAKDAY only (P=0.05). At the pen level, only TEMP and BUNK reflected illness. Importantly, steer-level illness variation was clearly related to degree of sickness response expression for all parameters.

Effects of short-term pre-calving handling of dairy heifers on heifers' behaviour and udder health

Silvia Ivemeyer[1], Marialuisa Pisani[1,2], Franziska Göbel[1] and Ute Knierim[1]
[1]University of Kassel, Farm Animal Behaviour and Husbandry Section, Nordbahnhofstraße 1a, 37213 Witzenhausen, Germany, [2]Università degli Studi della Basilicata, Dipartimento di Scienze delle Produzioni Animali, Viale dell'Ateneo Lucano 10, 85100 Potenza, Italy; ivemeyer@uni-kassel.de

The novelty of being handled during milking contributes to the peripartum stress in heifers with possibly adverse effects on their well-being and health. In the present study it was investigated whether pre-calving handling-training with circular or stroking touches led to reduced agitation of heifers during the sessions and during milking, reduced heifers' avoidance distances in the barn (AD) and improved udder health. On 3 farms, 4 training sessions (at 2 days with 1 day in-between and 2 per day) were conducted about 11 days pre-partum, touching hind quarters (HQ), hind legs and udder (HLU) and rest (dorsum, neck, flank, belly) of 13 trained heifers. Handling focus was on HLU. Sessions lasted about 10 min (average: 9.4 min ±2.4 min) and ended in a moment, when the heifer was calmer than at training start. 14 heifers served as control. Agitation responses (categorized by pre-defined behaviours into 'without aversive responses' and 'with aversive responses') of 10 heifers during the 4 handling-sessions (all heifers with complete video-taped trainings) were observed from videos and data analysed with general linear mixed models with handling number as fixed (time) effect and farm as random effect. The AD development from immediately before training to 3 to 4 days post-partum, agitation behaviour (stepping and kicking) during 2 milkings at 2 to 4 days post-partum, as well as somatic cell scores of the first three monthly milk test recordings (average and difference between first and third month) were analysed with mixed models with treatment as fixed and farm as random effect. Acceptable intra- and inter-observer reliability was achieved for the behavioural recordings. In the course of the 4 training-sessions, heifers' agitation decreased, indicated by increased endurance of touching HLU from the first (9.3% of total touching time) to the later trainings (second: 23.2%, third: 25.1%; fourth: 13.6%; time effect: P=0.004). Duration without aversive responses during touching HLU increased in the later trainings (second: 17.7%, third: 19.1%, fourth: 10.0% of HLU touching time) compared to the first training (6.0%; time effect: P=0.024). However, no significant differences between trained and control heifers were found regarding their behaviour after parturition and udder health (P>0.05). Possibly the handling was not intensive or long-lasting enough. Hence, conducted pre-calving handling had some positive effects on heifers' immediate agitation behaviour but no ongoing effects.

Long-term effects of dam-rearing: are there any benefits when heifers are introduced to the milking herd?

Tasja Kälber[1], Tolke Hechmann[2], Angelika Haeussermann[2], Susanne Waiblinger[3] and Kerstin Barth[1]

[1]*Thünen Institute of Organic Farming, Trenthorst 32, 23847 Westerau, Germany,* [2]*Institute of Agricultural Engineering, Christian-Albrechts-University, Max-Eyth-Straße 6, 24118 Kiel, Germany,* [3]*Institute of Animal Husbandry and Welfare, University of Veterinary Medicine, Veterinärplatz 1, 1210 Vienna, Austria; tasja.kaelber@ti.bund.de*

Dam-rearing until weaning seems to positively affect social skills. Our aim was to investigate if dam-reared heifers are better able to cope with being introduced into the milking herd. Nineteen dairy calves (German Holstein, GH; n=10; German Red Pied, GRP; n=9) were allocated to two treatments: rearing on automat feeder (A; n=7; 5GH; 2GRP) or reared with the dam (D; n=12; 5GH; 7GRP) for 90 d. Afterwards calves were reared under the same management. Approximately 27±3 d before calving, heifers were individually introduced into the milking herd. Cows were restrained in the feeding rack for 30 min, so the heifer could explore the barn undisturbed. Behaviour (feeding, movement, social behaviour, vocalization) was recorded continuously for the first twelve hours. In addition, heifers were equipped with an activity sensor to monitor lying, standing and walking behaviour for ten days. Statistical analyses were carried out using Wilcoxon tests (behavioural data) and linear mixed-effects models (data from sensor).No significant differences between treatments were found for latency or total duration of lying. Breed had an influence with GRP tending to lie earlier (in min., Median, Min-Max; GRP: 577, 31-720; GH: 720, 120-720; P=0.07) and total duration of lying (in s.,GRP: 18, 0- 603; GH: 0, 0-5; P=0.03). Only two out of seven heifers from A and four out of 12 from D lied within 12 h. No significant differences were found for feeding behaviour. Within the 12 h of observation all except 1 heifer (D) fed, but all heifers ruminated within the first 3 h. No significant differences were found for received aggressive behaviour or submissive behaviour. Only 2 heifers (one from each treatment) initiated social licking and three out of seven heifers from A and eight out of 12 from D never had socio-positive interactions. Frequency of vocalization tended to be higher in A within the first 30 min after introduction (A: 47, 13- 112; D: 16, 0-65; P=0.07), but declined dramatically in all animals afterwards without significant differences between groups. Length of standing bouts (median) increased from 30 s to 94 s in both groups after cows were released which was verified by the results from the activity sensor. Introducing heifers into the milking herd is stressful for the animals independent from rearing method, since heifers needed 5 days to adjust their activity pattern. Positive long-term effects of dam rearing related to introduction could not be confirmed during this study.

Agitation behaviour in the milking parlour during different attempts to stimulate milk ejection in cows rearing a calf or not

Katharina A. Zipp[1], Kerstin Barth[2] and Ute Knierim[1]
[1]University of Kassel, Farm Animal Behaviour and Husbandry Section, Nordbahnhofstr. 1a, 37213 Witzenhausen, Germany, [2]Thuenen-Institute of Organic Farming, Trenthorst 32, 23847 Westerau, Germany; zipp@uni-kassel.de

Dam rearing of dairy calves that can suckle at their mothers for several weeks after birth while the cows are also milked may have animal welfare advantages. However, disturbed alveolar milk ejection is one of the major challenges in this system. Stress in the suckled cows at milking may be one causal factor. In this study, suckled dairy cows (SC: n=15) and control cows, with calf separation half a day after birth (CC: n=21), in two experimental herds (one German Red Pied, one German Holstein) were exposed to one stimulus (olfactory=calf hair, tactile=teat massage, acoustic=recorded calf calls) per week between the 5th and 7th week of lactation. They were milked twice daily, and in each week there was also routine milking. One aim of this study was to assess the extent of possible stress responses to the different stimulations in suckled versus control cows in terms of agitation behaviour (stepping, kicking, elimination) in the parlour. This was videotaped and analysed in two phases: total time from entering to end of milking (TT), and from milking machine attachment until removal (MT). Data were analysed with mixed models for poisson (number of steps) and binomial distributed data (kicking and elimination: yes/no) and corrected for duration of TT and MT. The fixed effects, treatment (contrasts: stimuli vs. routine), group (SC, CC), breed, parity, all interactions and daytime were backwards selected. Random effects were animal and an ID to overcome overdispersion. Only significant factors remained in the fitted models. In general, in German Red Pied and primparous cows and at afternoon milkings more agitation occurred. There were indeed signs of increased stress in SC compared to CC, with elimination in 25% of milkings in SC and 8.4% in CC during TT (P<0.001). However, number of stepping and occurrence of kicking were not different. Both groups kicked more often during acoustic (TT: 29.8%, MT: 21.5%) and olfactory stimulation (TT: n.s., MT: 16.7%) than routine milking (TT: 23.6%, MT: 14.1%), while they kicked less during tactile stimulation (TT: 21.8%, MT: 12.9%, P<0.05). Treatment × group interactions indicated that during MT and acoustic stimulation CC stepped less and SC more than during routine milking (P<0.05). During olfactory stimulation SC but not CC showed decreased stepping (TT, MT: P<0.05) and elimination (TT: P<0.05) compared to routine milking. This might hint at a slight relaxing effect of the own calf's smell during milking, but further research is necessary to substantiate this finding.

Cumulative effect of environmental enrichment on behaviour and productivity of dairy goats kids after weaning

Erika Georgina Hernández, Andrés Ducoing, Julio César Cervantes and Anne María Sisto
FMVZ, Universidad Nacional Autónoma de México, UNAM, Circuito Exterior, Ciudad Universitaria, 04510, México DF, Mexico; gina_gato2003@yahoo.com.mx

The objective of this study was to evaluate the cumulative (since birth to the onset of puberty in contrast to enriching in each different stage) effect of environmental enrichment on behaviour and productive performance of weaned Alpine French female kids under confinement. Thirty-two female kids were randomly assigned from birth to two treatments (enriched (E) and non-enriched (NE)), with two replicates each (n=8). Physical enrichment was used, consisting in large wooden blocks to climb, brushes and hanging food (alfalfa, fruits and vegetables) The non enriched goats received the food in the feeder. Scan sampling was used every 5 minutes for 4 hours a day until completion of 150 hours of observation of individual time budgets from weaning (10 kg) until the onset of first estrous. The Kids were weighed every 15 days. Blood samples were taken to measure cortisol (levels of stress) and progesterone levels (onset of puberty). Cortisol and progesterone levels were measured using a commercial kit. A complete randomized design and multivariate analyses for repeated measures using contrasts to asses the effect of enrichment on the variables studied. The kids in the enriched group had a lower average proportion time showing agonistics encounters (P<0.01), exploration (P=0.05) and tended to spend less time in body care (P=0.07) than the non-enriched ones (E: 21.7±3.01, 38.7±5.1 and 35.1±3.05; NE: 3.6±3.01, 4.7±5.1 and 17.05±3.05, respectively). The difference in weight gain between the two groups was significant (P<0.05), the enriched group gained more weight (108.8±5.9 g) than the non-enriched one (87.9±5.9 g). Cortisol and progesterone levels did not show statistically significant differences (P>0.05). This study suggests that simple and low cost changes for environmental enrichment from birth have significant effects on the behaviour and weight gain of goat kids after weaning. Explorative behaviour of the non-enriched group was directed toward the enclosure, affecting facilities which can increase costs to the producer. The increase in agonistic encounters in the same group can be an indicator of poor welfare. Increase in weight gain in the enriched group may be decisive factor for adopting this practice.

Links between animals' seduction behaviour and technological artifacts

Pedro Bandeira Maia[1], Nuno Dias[2] and George Stilwell[3]
[1]*University of Aveiro and Polytechnic Institute of Coimbra, ESEC, ID+ Research Institute For Design, Media and Culture + ESEC, Arts and Technologies Department, Campus Universitário de Santiago, 3810-193 Aveiro, Portugal,* [2]*University of Aveiro, ID+ Research Institute For Design, Media and Culture, Campus Universitário de Santiago, 3810-193 Aveiro, Portugal,* [3]*Universidade de Lisboa, Faculdade de Medicina Veterinária, Avenida da Universidade Técnica, 1300-477 Lisboa, Portugal; pedrobandeiramaia@amadesign.net*

This research doesn´t intend to study animal behavior but uses what is already known and adapts it to artefacts, promoting an association between 2 distinct areas, attempting to solve some of existing problems in society. It is expected that this approach based on the seduction behaviors founded in nature, may generate a new paradigm in human relationship with material goods that facilitate communication between people. The strategy of building this taxonomy began with a survey of examples of rituals of seduction, which we believed to have visual, poetic, rhythmic and conceptual ability for application in design. We define 5 categories of behaviors (aggressive, gentle behavior, tricky, ritualized and materials), 3 groups of biological agents (flying, terrestrial and aquatic) and 4 time periods (fast, short, medium and long). Finally were defined some characteristics observed in the process, such as kindness, courtesy, intimidation, etc. Based on this classification, we developed a framework of rituals with potential for application in design. Later we started visual representation of the selected rituals by designing infography in order to be able to transform an 'abstract' information into a visual one based in conceptual and poetic representations that increase imagination in design. Through the developed Taxonomy we started to implement these rituals in products, strategies and services. At this stage we are working with people who have difficulty in relating to others, either for psychological/behavioral issues or they have some kind of disability. We started the development of a jewerly collection where their shape, color, odor or dynamic are inspired by the rituals of seduction and with the incorporate technology are able to promote communication between people with some kind of sensory or behavioral handicap. Objects like earphones that act as hearing aids, are jewerly but at the same time promote a ritualized interaction with others who use the same headset via a unique code. Another ongoing project are cellphone covers based on Chameleons rituals that establish patterns and colors, that without resorting to words, start a game of interaction between two persons in order to establish a game of seduction and seek strategies to circumvent the problem of shyness loving studied in Psychology.

Carcass skin lesions in boars reflect their aggressiveness on farm

Dayane Teixeira[1], Nienke Van Staaveren[1], Alison Hanlon[2], Niamh O'connell[3] and Laura Boyle[1]
[1]Teagasc Moorepark, Pig Development Department, Fermoy, Co. Cork, Ireland, [2]University College Dublin, UCD Vet Sciences Centre, Belfield, Dublin 4, Ireland, [3]Queens University Belfast, School of Biological Sciences, Medical Biology Centre, 97 Lisburn Road, Belfast BT9 7B, Northern Ireland, United Kingdom; dayane.teixeira@teagasc.ie

The aim was to investigate if aggressive and mounting (sexual) behaviours performed by boars are associated with skin lesions scored on farm and at the abattoir compared to gilts. 141 pigs were accommodated in 5 single-sex groups per gender (14 pigs/group) on fully slatted pens. Data collection took place for 2 weeks prior to slaughter. Posture (lying, sitting and standing), aggression (head knock and fight) and mounting behaviours were recorded in 3×2 hour periods on days -13, -9, -7 and -2. On days -14 and -1 pigs were weighed and the ears, neck, shoulders, side, belly, back and hindquarters were scored according to the sum of severity of each lesion (0-3). Also, all scores were summed to provide a total lesion score. Carcass cold weights were obtained from the abattoir. In the chill room, skin lesions were assessed as per Welfare Quality® protocol (WQ) and bruises were counted and classified as fighting, mounting and handling-type. Data were analysed in SAS (weight and behaviour: PROC MIXED; lesion scores: PROC NPAR1WAY; correlations: PROC CORR). Boars were heavier than gilts (100.7 vs. 99.0 kg; s.e.m. 0.59; $P<0.05$) but there was no effect of gender on carcass cold weight (77.3±0.4 kg). Boars performed more aggressive (1.8 vs. 1.0 aggression/pig; s.e.m. 0.22; $P<0.05$) and mounting behaviours (0.4 vs. 0.005 mounts/pig; s.e.m. 0.02; $P<0.05$) than gilts. Lying active (5.0±0.8%), lying inactive (64.9±2.8%) and sitting (6.8±0.6%) behaviours were similar in both genders, as well as time standing at period 8-10 h (31.6±2.9%) and 14-16 h (20.5±2.9%). But boars spent more time standing between 11-13 h than gilts (21.8 vs. 13.7%; s.e.m. 2.86; $P<0.05$). On Day -14, total skin lesion scores were similar in both genders, but on day -1 boars had higher scores than gilts (11.2 vs. 8.2; s.e.m. 0.95; $P<0.05$). Boars had higher total skin lesion scores on the carcasses (1.9 vs. 1.3; s.e.m. 0.10; $P<0.05$) and more fighting-type bruises (4.5 vs. 2.3; s.e.m. 0.35; $P<0.05$) than gilts. There was no association between aggressive behaviour and skin lesions scored on farm on Day -1 (actor: $r=0.004$, $P>0.05$; recipient: $r=0.022$, $P>0.05$) but there were positive correlations between aggressive behaviour and skin lesions scored on the carcass (actor: $r=0.383$, $P<0.0001$; recipient: $r=0.294$, $P<0.001$, respectively) and fighting-type bruises (actor: $r=0.442$, $P<0.0001$; recipient: $r=0.297$, $P<0.001$, respectively). There was no association between skin lesion scores recorded on farm and on carcass ($r=-0.093$, $P>0.05$). Carcass skin lesions were a more sensitive indicator of injury and poor welfare caused by the aggressiveness of boars. Boar sexual behaviour was not reflected in the lesion pattern of the skin probably due the low frequency during the study.

Case study: the effects of the constituent factors of grazing system on welfare of fattening pigs

Akitsu Tozawa, Shigefumi Tanaka and Shusuke Sato
Tohoku University, Graduate School of Agricultural Science, 232-3, Yomogita, Naruko-Onsen, Osaki, Miyagi 989-6711, Japan; akitsu-t@bios.tohoku.ac.jp

To clarify the most effective constituent factor of grazing (outdoor pasturing system) for improving welfare of fattening pigs, 20 pigs were equally allocated to five treatments: Indoor housing system (IS), Outdoor pasturing system (OP), Concrete floor paddock system (CF), Concrete floor paddock system with fresh grass (FG), and Soil floor paddock system (SF). There was a pen with a roof in open air (1.8×2.7 m) in each treatment and a paddock (10×20 m) except IS. Behaviours and productivities as average daily gain (ADG) and feed conversion ratio (FCR) were observed in the prior period (100-124 days old) and the latter period (138-164 days old). Wounds on the body were observed before shipping. Data were analysed by one-way ANOVA between the treatments and t-test between the periods. CF pigs behaved as IS pigs, except they expressed less disturbed behaviour such as tail-biting, inanimate-biting and sham-chewing than IS pigs in the prior period (CF 1.6±1.7, IS 9.0±6.0 times/h: $P<0.05$). FG pigs foraged (34.5±3.9%) and chewed (15.9±4.9 times/h) the most in the prior and became inactive (resting: 70.4±9.6%) in the latter. SF pigs as OP, expressed rooting (Prior 22.0±4.6, Latter 27.3±3.6 times/h) and playing (Prior 3.4±1.8, Latter 1.6±0.8 times/h) more than pigs in other treatments ($P<0.05$). Productivities, such as ADG(kg/day) in the latter, CF (0.72±0.09) were similar to IS (0.67±0.15), FG (0.52±0.16) were the lowest, and SF (0.89±0.24) tended to increase more than the prior (0.59±0.14, $P<0.1$) as OP (0.91±0.09). It is concluded that increasing space allowance and the opportunities to express chewing are not enough for improving welfare of fattening pigs and the existence of a soil floor is the most important constituent factor of grazing.

Effect of increased room temperature and a new heat source on piglets' use of the heated creep area

Mona Lilian Vestbjerg Larsen, Lene Juul Pedersen and Karen Thodberg
Aarhus University, Dept. of Animal Science, Blichers Allé 20, 8830 Tjele, Denmark;
mona@science.au.dk

Hypothermia, experienced by neonatal piglets, has been related to piglet death, especially within the first days of life. Use of the heated creep area is low during these days. It is therefore important to improve both the creep area and the thermal environment outside of the creep area. We wanted to test if piglets' use of the creep area was (1) reduced by increasing the room temperature in the farrowing unit and (2) increased by installing a new heat source (Eheater) in the creep area with a more even, widespread heat surface. Twenty crated sows were randomly assigned to two room temperatures during the periparturient period: 20 °C (CONT, n=10)) or 25 °C (HEAT, n=10) until 12 hours after last farrowing, and to two heat sources in the creep areas: a standard infrared heat lamp (STAND, n=10) or a newly invented radiant heat source (EHEAT, n=10), with five of each heat source within the two room temperatures. Observations on piglets' use of the creep area were made as scan sampling every ten minutes in a four hour time period on day 1-7, 14, and 21 after farrowing. To diagnose fever (>39 °C) in sows, their rectal temperature was measured each day. For statistical analysis, a linear mixed model was used. A three-way interaction between room temperature, heat source and days after farrowing ($F_{6,95}=2.20$; P=0.05) was seen during the first 7 days. The EHeater particularly assured an earlier use of the creep area at low room temperature during the 2nd and 3rd day, with 40% of the piglets using the creep area compared to 17% for STAND litters and to 15% and 0% for EHEAT and STAND litters, respectively, at a high room temperature on the 2nd day. Fever diagnosed in sows also had an effect on piglets' use of the creep area ($F_{1,94}=7.27$; P=0.008) with less use if fever was present, indicating that the thermal microclimate in the creep area might not be the only explaining factor when low use of the creep area is observed. The results indicate that both room temperature and the new heat source have an impact on piglets' use of the creep area, especially during the first days of life where piglets otherwise tend to stay at the sows udder.

Effect of drinker position on lying and excreting behaviour of growing-finishing pigs in a welfare friendly housing facility

Marko Ocepek[1,2,3], Mirjana Busančić[2,3], Vilma Šuštar[2], Dejan Škorjanc[2] and André J.A. Aarnink[3]
[1]Norwegian University of Life Sciences, Animal and Aquacultural Sciences, P.O. Box 5003, 1432 Ås, Norway, [2]University of Maribor, Faculty of Agriculture and Life Sciences, Pivola 10, 2311 Hoče, Slovenia, [3]Wageningen University and Research Centre, Livestock Research, P.O. Box 135, 6700 AC Wageningen, the Netherlands; marko.ocepek@nmbu.no

The concept of the innovating pig housing system 'Star+' is to ensure optimum conditions with respect to animal welfare and the environment. By providing, indoor and outdoor areas, special inlets for daylight, a natural ventilation system, system for continuously manure removing, the reduction of energy costs as well as a clean and fresh environment for pigs and human are ensured. Pens have a large solid floor area so pigs can lie on a solid floor during the whole growing period. Such a large solid floor is very vulnerable for fouling. It was questioned whether this fouling could be influenced by the position of the drinkers. Therefore, in two rounds (October till January; February till June) we studied the impact of treatment (two drinkers indoor (IN), two drinkers outdoor (OUT), and one drinker indoor/one drinker outdoor (IN_OUT)) on the lying/excreting preferences of the pigs on solid and slatted floors. Pigs were assigned to one of the 12 pens (4 per treatment) and housed in groups of 18 (9 boars + 9 sows) per pen. Each pen consisted of an indoor area of 15.9 m^2, and an outdoor area of 5.7 m^2. The indoor pen floor consisted of 75% concrete solid floor, and 25% metal triangular slatted floor. In the outdoor area, the floor consisted of 35% metal triangular slatted floor and 65% concrete slatted floor. Observations of pigs' lying preferences and urination/defecation preferences were performed every 2 weeks (24 h) and analysed using mixed effects analysis in R. For excretions on the solid floor, results showed urination frequencies of 40, 54, and 48% ($P<0.001$) and defecation frequencies of 26, 33, and 33% ($P<0.001$) for OUT, IN, and IN_OUT, respectively. No difference was found between treatments in number of urination on indoor slatted floor, while a tendency of less defecation was observed in IN (1.9%) and IN_OUT (1.4%) treatment compared to OUT treatment (2.4%), but this difference was not statistically significant. Higher urination frequencies outdoor (55 vs. 42 vs. 47%; $P<0.001$) were found in OUT compared to IN and IN_OUT treatments, respectively. The results show that position of the drinkers influences pig behaviour. By placing the drinkers in the outdoor area pigs urinate and defecate more frequently in outdoor slatted floor areas and less on indoor solid floor areas. This is favourable for lower ammonia emissions, less time needed for manual cleaning, and a cleaner solid floor area for lying.

Influence of prenatal experience with human voice on the neonatal behavioural reaction to human voices with different emotion

Céline Tallet[1], Marine Rakotomahandry[1], Carole Guérin[1], Alban Lemasson[2] and Martine Hausberger[2]
[1]*INRA UMR1348 PEGASE, domaine de la prise, 35590 Saint-Gilles, France,* [2]*UMR 6552 CNRS, Université de Rennes 1, Laboratoire d'éthologie animale et humaine, 35380 Paimpont, France;* celine.tallet@rennes.inra.fr

Farm animals have to adapt to human presence from birth and being handled may lead to fear and stress reactions. It is known that the mother can be used as a postnatal model in the development of young-human relationship. Through her, some information like auditory ones may even be learnt prenatally. We tested this idea in pigs because they communicate a lot by acoustic signals. The hypotheses were that prenatal experience with human voice could modify behavioural reactions to the experienced voice and to an unfamiliar voice expressing different emotions. We worked with 30 pregnant sows from the last month of gestation. Ten sows (treatment A) were submitted to recordings of human voices during handling: vA during positive interactions and vB during negative interactions, twice a day, 5 days a week, for 10 minutes. Ten other sows (treatment B) received the contrary, i.e. vB during positive interactions and vA during negative interactions. Ten last sows (treatment C) received no vocal stimulations during handling sessions. Two days old piglets (36 A, 39 B, 35 C) were submitted to a 5 min choice test between voices vA and vB in a testing pen (2×1 m). Each voice was played back through loudspeakers positioned at each end of the pen. At 15-18 days of age, 20 other piglets from each treatment were tested in the same conditions except that we played back the voice of an unknown person, reading the same text with a joyful or angry intention. In both tests we recorded vocalisations and locomotion. Data were analysed using non parametric statistics (Statview). In both tests, A and B piglets started to move sooner (P<0.01) and produced less stress related vocalisations than C piglets (P<0.05). All piglets spent more time close to the loudspeakers (<50 cm) than at a distance (P<0.001). The latency (median (IQ): 54 s (86 s), P>0.05) to be and the time spent (178 s (46 s)) close to the loudspeakers did not depend on the treatment (P>0.05). We also found no difference between the time spent close to one loudspeaker or the other, neither for vA versus vB, nor for joyful versus angry intention (P>0.05). The results show that the prenatal experience of human voice reduces postnatal behavioural reactions of stress (vocalisations, latency to move) during the playback human voices. However, it does not seem to induce specific attraction toward human voice, or human emotional intention. Therefore prenatal experience with human voice may be a good way of reducing fear reactions to human voice after birth.

Fresh wood, plastic pipe or metal chain – which objects reduce tail and ear biting on commercial pig farms?

Helena Telkänranta[1], Marc Bracke[2] and Anna Valros[1]
[1]University of Helsinki, Department of Production Animal Medicine, P.O. Box 57, 00014 University of Helsinki, Finland, [2]Wageningen UR Livestock Research, P.O. Box 65, 8200AB Lelystad, the Netherlands; helena.telkanranta@helsinki.fi

Our aim was to test three low-cost point-source manipulable objects for pigs to see whether they sustain pigs' interest for three months and yield welfare benefits. Two Finnish commercial farms with undocked pigs and partially slatted floors were used. Experiment 1 included 780 growing-finishing pigs from the age of 2 months to slaughter. Control pens (C, n=17), and all treatment pens, had a straw rack, a metal chain and wood shavings. Treatment pens had additional objects: horizontally suspended pieces of fresh wood, 30 cm/pig (W, n=14); a cross of two 60-cm polythene pipes (P, n=13); two crosses of metal chains (B, n=15); or all of the above (WPB, n=14). After 3 months, tail and ear damage were scored and the pens were video-recorded. Behaviour was scored during a 20-minute period. Experiment 2 included 656 growing-finishing pigs from weaning to slaughter. The control pens (C, n=16), and all treatment pens, had a straw rack. Treatment pens also had one of the following, suspended horizontally, 10 cm/pig: pieces of birch wood (W, n=16), or polythene pipe (P, n=12). After 4 months, tail and ear damage were scored, and the latency to approach an unfamiliar person (LAP) was recorded. Treatment effects were tested with one-way ANOVA (SPSS 21). In Experiment 1, the prevalence of undamaged tails was higher in W and WPB (mean 55% in both) than in C (34%; $P<0.05$). Pigs manipulated objects more frequently in W, P and WPB (2.7, 2.9 and 3.1 manipulations/pig/20 min) than in B (0.8) and C (0.5) ($P<0.001$). In Experiment 2, W had less ear damage, using a 5-point scale (1.3) than P and C (1.8 and 1.9; $P<0.01$). LAP was shorter in W (8.8 s) than in P and C (17.3 s and 15.4 s, $P<0.05$). Horizontally suspended pieces of fresh wood can improve welfare in commercial pig farming.

Behavioural development of piglets in farrowing crates and in a multi-suckling system

Sofie E. Van Nieuwamerongen, Annemarije Verspuy, Nicoline M. Soede and J. Elizabeth Bolhuis
Wageningen University, Dept. of Animal Sciences, De Elst 1, 6708 WD Wageningen, the
Netherlands; sofie.vannieuwamerongen@wur.nl

Commercial use of group housing systems for lactating sows is limited, but the recent transition to group housing during gestation in the EU may result in a renewed interest in such systems. Multi-suckling (MS) systems, for instance, in which sows are grouped together with their litters, may provide a better transition from gestation housing. Additionally, MS housing may benefit piglet development compared with conventional housing in farrowing crates (FC), by stimulating play behaviour and, possibly, by reducing harmful oral manipulative behaviour directed at pen mates, e.g. tail and ear biting. In addition, piglet feed intake may be stimulated, especially when social learning of foraging behaviour is facilitated by use of a communal feeding area. Adaptation to solid feed during the suckling phase is important for a successful transition to the post-weaning period. One week before farrowing, 20 multiparous sows were allocated to the MS system or to farrowing crates (2 batches with 5 sows per system). MS sows farrowed in individual pens with 2 kg of straw and had access to a communal area (with feeding area) with 5 ropes and 5 hessian sacks throughout lactation. The piglets gained access from 1 week of age. Of 4 piglets per litter, frequencies of oral manipulation of sows and other piglets, play behaviour and foraging-related behaviour were recorded using 4×10 min. continuous behaviour sampling per week at 2, 3 and 4 weeks of age. Effects of housing system, week and their interaction were analysed with mixed models with batch nested within system as random effect, i.e. a replicate of 5 sows was the experimental unit. MS piglets played 2.6 times more often (7.0 ± 0.3 vs. 2.7 ± 0.5, $P<0.01$) and showed 2.4 times less manipulative behaviour (1.4 ± 0.3 vs. 3.4 ± 0.1, $P<0.05$) than FC piglets. Play and manipulative behaviour were unaffected by week. Frequency of foraging-related behaviour (i.e. sniffing, nosing or eating sow or piglet feed) did not differ between the two systems (3.5 ± 0.7), but tended to increase with age (week effect $P<0.10$). In conclusion, although MS piglets had more opportunities to eat together with sows and other piglets, the frequency of foraging-related behaviour did not differ from that of FC piglets. The MS system promoted piglet play behaviour and reduced harmful manipulative behaviour, compared with FC housing, likely due to the physically and socially enriched and more spacious environment. The reduced manipulative behaviour may decrease the risk of tail and ear biting problems in later life. Future research in the MS system will address long-term consequences of these developmental differences on pig performance.

Effects of alternative farrowing systems on piglet mortality and behavior

Laurie A Mack[1], Shawna P Rossini[1], Sabrina J Leventhal[2] and Thomas D Parsons[1]
[1]University of Pennsylvania, School of Veterinary Medicine, 382 W. Street Rd., Kennett Square, PA 19348, USA, [2]University of Delaware, 531 S. College Ave., Newark, DE 19716, USA; lauriem@vet.penn.edu

In the United States, commercial sows are often housed in crates that limit locomotion and natural behavior during gestation and lactation. Sow welfare and consumer concerns are motivating changes in gestational sow housing. During lactation, farrowing crates present similar welfare challenges as gestation crates for the sow. However, lactational housing also affects the pigs whose welfare may diverge from that of the sow in a given system. A better understanding of various farrowing systems could benefit both sow and pigs. Sows and litters were housed in either: (1) a closed farrowing crate, 1.0 m^2/sow (n=19); (2) a hinged crate that when opened 2 wk after parturition provided turn-around space for the sow, 1.6 m^2/sow (n=19); or (3) a thinly bedded pen, 6.0 m^2/sow (n=20). Deaths were recorded until weaning at 5 wk of age. At 27.26±0.59 d of age, 2 female and 2 male pigs per litter were separately observed for 5 min in a 3.24 m^2 arena containing a novel food, a cookie (isolation test). Another 2 males and females per litter were observed in the arena with an unfamiliar sex-, age-, and treatment-matched pig (social test). On the following day, each pig was tested in the opposite test. Data were analyzed using a mixed model with treatment, sex, test order (for behaviors), and their interactions as fixed effects and pig nested in sow as a random effect (PROC GLIMMIX, SAS 9.3). At 26.0%, penned litters had greater mortality than closed (10.3%, P<0.001) or hinged (15.2%, P=0.005) crated litters. Housing had no effect on the piglets' cookie eating behavior in the isolation test or body posture, distance travelled, or location in either test. However, in the isolation test on test d 1, pigs from closed crates excreted more (P=0.02) and those from the hinged crates tended to excrete more (P=0.08) than penned pigs, but on d 2 the treatments did not differ. In the social test over both days, penned pigs fought longer than closed crated pigs (P=0.04) with those from hinged crates intermediate. Male pigs from pens spent more time in non-aggressive contact and less time in non-social behavior than those from closed (P=0.027 and <0.001, respectively) and hinged crates (P<0.001, both), but housing did not affect the female pigs. In summary, housing litters in pens increased piglet mortality and altered behavior. Penned pigs of both sexes fought longer, but penned males also exhibited more non-aggressive social behavior. This suggests that alternative farrowing systems can have complex effects on piglet welfare that need careful evaluation.

Skin lesions in entire male pigs in relation to aggressive and mounting behaviour in response to mixing prior to slaughter

Nienke Van Staaveren[1,2], Dayane Teixeira[2], Alison Hanlon[1] and Laura Boyle[2]
[1]*School of Veterinary Medicine, UCD, Dublin, Ireland,* [2]*Teagasc, Pig Development Department, Moorepark, Ireland; nienke.vanstaaveren@teagasc.ie*

The aim of this study was to determine the impact of mixing entire male pigs prior to transport to slaughter on aggressive and mounting behaviour and the extent to which these behaviours are reflected in skin lesions and loin bruising. Pigs from 3 single sex pens (30.8±6.4 pigs/pen) were randomly allocated to one of three treatments (n=20/group; 6 focal pigs/group) on the day of slaughter over 5 replicates: males unmixed (MUM); males mixed (MM); and males mixed with females (MF). Pigs were mixed prior to transport in the holding area on the farm (1.35±0.37 m^2/pig). All occurrence behaviour sampling was used to record the frequency of aggressive and sexual behaviours during holding on farm before loading (approx. 1 h) and at lairage (0.65±0.03 m^2/pig, approx. 1 h). The actor and recipient were recorded when focal pigs were involved. Skin lesions on the left side of the body were scored on a 0-5 scale at lairage (focal pigs) and on the carcass (all pigs). Carcass loins were assessed for bruising according to severity (0-2). Data were checked for normality and effects of treatment and time on behaviour were analysed by SAS V9.3 PROC MIXED. Correlations between behaviour and skin lesion/loin bruising scores were calculated (PROC CORR). MM performed significantly more aggressive behaviours than MUM (50.4±10.72 vs. 20.3±9.55; P<0.05) and tended to perform more aggressive behaviours than MF groups (37.2±10.77; P=0.06). MM performed significantly more mounting behaviour than MF and MUM groups (30.9±9.99 vs. 11.4±3.76 and 9.8±3.74 respectively; P<0.05). More aggressive (56.3±9.61 vs. 19.7±5.42) and mounting (29.3±7.61 vs. 7.9±2.18) behaviours were observed during holding on the farm than at lairage (P<0.01). Pigs that were involved in a higher number of fights after mixing had a higher skin lesion score at lairage (r=0.23; P<0.05) but there was no relationship between aggressive behaviour and carcass lesion scores (P>0.05). There was no relationship between mounting behaviour and skin lesion/loin bruising scores (P>0.05). Mixing entire males prior to transport stimulates mounting and aggressive behaviour particularly if mixed with other males. Aggressive behaviour was reflected in skin damage on the live animal but not on the carcass, possibly due to damage to the carcass caused by pre-slaughter handling and post-slaughter processing such as scalding. The increase in mounting behaviour stimulated by mixing was not reflected in higher skin lesion or loin bruising scores. These findings have important implications for the development of measurements relating to lesions/bruises on the carcass as welfare indicators.

The impact of umbilical outpouchings on the behavior of slaughter pigs housed in a pick-up facility

Sarah-Lina Aa. Schild[1], Tine Rousing[1], Henrik E. Jensen[2], Kristiane Barington[2] and Mette S. Herskin[1]
[1]Aarhus University, Department of Animal Science, AU-Foulum, P.O. Box 50, 8830 Tjele, Denmark, [2]University of Copenhagen, Department of Veterinary Disease Biology, Copenhagen, Denmark; sarahlina.schild@gmail.com

In Danish pig production, finisher pigs are moved to a pick-up facility before being transported to slaughter. The aim of the present study was to examine behavioural and clinical consequences of a 6 h stay in a model pick-up facility (no access to feed or enrichment, 0.65 m^2/pig and mixed with unfamiliar pigs at entry) in pigs with umbilical outpouchings (UOs) (diagnosed post mortem as either UH (umbilical hernia) or NHUO (non-hernia umbilical outpouching)) and healthy pigs in order to gain knowledge about effects of this relatively common clinical condition. This study comprised 28 pairs of slaughter pigs, each pair consisting of one UO-pig and one healthy control. The size of the UOs varied from 5 to 20 cm in diameter with a median diameter of 12 cm. All experimental animals were subjected to a clinical evaluation, including scoring of skin lesions, before being transferred to the model pick-up facility. The behaviour of the pigs was observed during the 6 h stay, after which the pigs were subjected to a second clinical evaluation. At the end of the study, all animals were euthanized and pigs with UOs were examined pathologically. The stay in the pick-up facility led to a general increase in score of skin lesions (before: 0; after: 3; P<0.001). Several significant differences in the behaviour of UO-pigs compared to healthy controls were found. UH-pigs differed behaviourally from pigs NHUO- and control pigs. The proportion of lying was lower in UH-pigs compared to NHUO- and control pigs (79±2, 85±1 and 83±1, respectively; P=0.01) and UH-pigs showed higher proportion of time spent sitting than both NHUO-pigs and controls (5±1, 3±1, 3±0.3, respectively; P<0.005). Unexpectedly no significant size-related effects of the UOs were found. Irrespectively of the presence of the UOs, the stay in the pick-up facility led to increased lesion scores, suggesting that this site in the production chain may challenge animal welfare. The behaviour of pigs with UO, and especially UOs diagnosed as hernia, was influenced by the stay in the pick-up facility indicating that the presence of UH might reduce the ability of the animals to cope with the stay in a typical pick-up facility.

Rearing environment affects preference for sham dustbathing on different types of scratch mats

Sophie Taylor, Eugenia Herwig, Michelle Hunniford and Tina Widowski

University of Guelph, Animal & Poultry Science, 50 Stone Rd E, N1G 2W1 ON, Canada; twidowsk@uoguelph.ca

Furnished cages for laying hens are outfitted with scratch mats that are intended to support dustbathing behaviour (DB), to some degree, and many new designs of cages include scratch mats made of a variety of textures and sizes. Artificial turf has been previously investigated as a substrate for (sham) DB in cages, but it is not known whether hens' perceive mats made from other materials as suitable DB substrates. Our objectives were to determine the preferences of laying hens to sham DB on different types of scratch mats and whether previous experience affected their preferences. We tested 36 24-week old LSL-Lite laying hens reared from day 1 and housed at 16 weeks in either in aviaries with access to wood shavings from 7 weeks (n=18) or in conventional cages with wire floors (n=18). Each hen was tested individually over 3 days in a large cage that had 3 mats (each 20×20 cm) made of different materials fixed to the wire floor (WR) directly in front of the feed trough: plastic artificial turf (AT), smooth plastic (SM) and textured plastic (TX). Prior to testing, all hens were housed in their respective rearing groups in a large cage for 1 week with a 2-day exposure to each of the mats. A single observer blinded to treatment scored the video records from 1000-1545 h each day of the test to identify duration and location of sham DB bouts. Data were summed for the 3-day test, and proportions of time spent sham DB on each surface were transformed and analyzed using mixed model ANOVA (SAS 9.3). All 18 Cage-reared hens sham-DB during the trial, while half (n=9) of the Aviary hens did not. The average total duration DB on all surfaces was 326.7±162.0 s for Aviary hens and 696.3±110.9 s for Cage. Of the hens that sham DB, those from the aviary performed a greater proportion on AT than those from cages (0.37±0.11 vs 0.08±0.02; F=8.26, P=0.009). Cage hens performed a greater proportion of DB on WR compared to Aviary (0.70±0.05 vs 0.48±0.13, F=3.97; P=0.057). There were no differences between rearing groups on the proportion of DB performed on SM (aviary: 0.05±0.02, cage: 0.17±0.06; F=0.14, P=0.711) or TX (aviary: 0.10±0.12, cage: 0.05±0.02, F=0.88, P=0.36). This is the first study to directly compare the effects of rearing environment on preference for scratch mats for furnished cages. Our results explain some of the discrepancies in the literature where AT has been preferred over WR in experiments using hens reared on litter but not in experiments using hens reared on WR. Our results also indicate that SM and TX mats were not attractive substrates for sham DB regardless of rearing experience.

The impact of phenotypic appearance on stress and immune responses in laying hens: Is it a group size dependent phenomena?

Raul H Marin[1,2,3], Franco N Nazar[1], Guiomar Liste[3], Irene Campderrich[3] and Inma Estevez[2,3]
[1]IIByT, CONICET and UNC, Vélez Sarsfield 1611, 5016 Córdoba, Argentina, [2]IKERBASQUE, Basque Foundation for Science, Alameda Urquijo 36-5, 48011 Bilbao, Spain, [3]Neiker-Tecnalia, Animal Production, P.O. Box 46, 01080 Vitoria-Gasteiz, Spain; raulmarin1@hotmail.com

Alteration of birds' phenotypic appearance (PA) may lead to unwanted behaviours, potentially impairing poultry welfare, health and productivity. Likewise, group size (GS) may play an important role modulating the expression of adaptive behaviours and stress response. This study evaluated whether manipulation of the PA and GS in Hy-line Brown laying hens may affect stress and immune responses. At 1 day of age, 750 chicks were randomly assigned to 30 pens at GS either 10 or 40 (8 hens/m^2). At arrival, PA of 0, 30, 50, 70 or 100% of the birds in each pen were artificially altered by black dying their heads back (Marked-M). Remaining birds were unchanged (Unmarked-UM). At 32 weeks, basal and acute stress adrenocortical response, heterophil/lymphocyte (H/L) ratio, lymphoproliferative response to phytohemagglutinin-p and primary antibody response against sheep red blood cells were measured in 6 birds/pen (3 M and 3 UM within the PA heterogeneous pens and 6 M or 6 UM from homogeneous pens). ANOVAs showed no differences among treatment combinations. In a second phase, birds in initially homogeneous pens (0 and 100%), were either M or UM sequentially to reach 30, 50 and 70% of the hens altered at 34, 38 and 44 weeks, respectively. Initially heterogeneous pens remained unaltered and were used as controls. Two weeks after last PA manipulation, mentioned variables were measured again. Both altered and non altered hens within altered pens showed increased (P<0.01) H/L ratios compared to their unchanged control counterparts suggesting that the appearance of new phenotypes triggered a social chronic reaction in all pen members whether altered or unaltered. After a social isolation test, all groups showed increased (P<0.01) corticosterone responses. However, within GS 40, the hens that remained with their PA unaltered evidenced a significantly lower (P<0.05) response than their PA altered groupmates. This suggest that in GS 40, unaltered hens within altered pens were able to better cope when exposed to a new social stressor probably due to previous experience that may potentially enhance social plasticity favored by social interactions emerging after the PA alterations of their penmate counterparts. Although all birds in altered pens showed modified blood cells, their antibody and lymphoproliferative responses did not differ from their control counterparts suggesting that all groupmates were able to immunologically cope with the chronic social stress induced at least within the time frame evaluated.

Do broiler breeders prefer elevated sleeping sites?

Sabine G. Gebhardt-Henrich[1] and Hans Oester[2]
[1]University of Bern, VPH Institute, Dept. Animal Welfare, Burgerweg 22, 3052 Zollikofen, Switzerland, [2]ZTHZ, Mezenerweg 3, 3013 Bern, Switzerland; sabine.gebhardt@vetsuisse.unibe.ch

Unlike laying hens, broiler breeders are commonly kept without perches although perches are mandatory in the Swiss animal welfare ordinance. To examine the potential need for perches in broilers we conducted two preliminary studies to assess whether fast (Ross) and more slowly growing (JA) broiler breeders (1) used elevated structures such as elevated manure slats or the grill above the feeders at night and, (2) whether they would use a low perch. In the first study, four 3-m sections in a 60 m long barn with Ross and a 60 m long barn with JA were marked and 2 pictures were taken per section at 20:20 hours ten times between 5-53 weeks. In the second study four sections in a barn with Ross broiler breeders were marked while two sections were equipped with 45 m plastic perches along and perpendicular to the nest directly on the manure slats after rearing. Pictures were taken when the birds were 25, 45, and 60 weeks old. The number of birds on the manure slats, the grill above the feed trough, and perches (if present) were counted and divided by the total number. Arcsine transformed data were analyzed with GLM with section as subject and age and time as repeated factors. For study 1: JA hybrids used elevated structures earlier ($49\pm21\%$ of the birds at 10 weeks of age) and to a greater extent ($91\pm5\%$ of the birds after 20 weeks of age) than Ross ($20\pm10\%$ at 10 weeks of age, $80\pm5\%$ after 20 weeks of age) (hybrid: $F_{1,28}=27.64$, P<0.0001, age: $F_{1,42}=38.72$, P<0.0001). JA hybrids continued using elevated structures in high numbers whereas Ross used these structures less when they got older (interaction hybrid*age: $F_{1,30}=14.16$, P<0.0004). Study 2: The age effect was the same as in Study 1. Fewer than 1% of the birds used the perch and the perch had no effect on the use of other elevated structures (age: $F_{1,39}=442.96$, P<0.0001, perch: $F_{1,2}=6.34$, P=0.13). In summary, broiler breeders used elevated structures such as the grill above feed troughs and the edges of the elevated manure slats at night. Perches were rarely used, possibly because they were lower than the grill above feed troughs. The reason for the decrease in use of elevated structures with age in Ross hybrids is unclear though the structures might have been unsuitable for heavy birds. Future study will need to examine the use of perches at different heights and aviaries.

Relationship between lameness and body weight in broiler chicken

Bruno Roberto Müller, Bruna Nascimento Tulio, Ana Paula de Oliveira Souza and Carla Forte Maiolino Molento

Universidade Federal do Paraná, Zootecnia, Rua dos Funcionários, 1540, Juvevê, Curitiba-PR, 80035-050, Brazil; brunormuller@yahoo.com.br

Producing heavier animals is considered advantageous in terms of productivity and profitability. This strategy, on the other hand, leads to an unbalanced situation, where monetary gains are prioritized in detriment of animal welfare. Lameness constitutes an important welfare issue for broiler chickens, and is often related to the rapid growth rate of the modern artificial lineages. The objective of the present work was to evaluate the relationship between walking ability and body weight of 200 Cobb® broilers and to study the effects of this interaction on productivity. Male birds were selected from two flocks housed in standard acclimatized aviaries in southern Brazil, at 43 days of age, and distributed into four groups according to their gait scores (2, 3, 4 and 5) until each group contained a total of 50 animals. Gait scores were defined according to an established method, where score 0 reflects an animal with no degree of impairment and score 5 indicates birds that cannot walk at all. Animals were then weighed and returned to the flock. Analysis of variance revealed significant differences ($P<0.05$) on body weight between birds with different gait scores, where birds with scores 2 and 3 were equally heavier (3.02 ± 0.21 kg and 3.17 ± 0.20 kg, respectively) than birds with scores 4 (2.59 ± 0.47 kg) and 5 (2.11 ± 0.50 kg). Regression analysis showed a cubic interaction ($P<0.01$) of body weight and walking ability, indicating that lameness is a multifactorial problem and might be expressed in two concurrent scenarios: as a consequence of rapid growth rate and exaggerated weight combined with weak bone structure, or as an early problem which leads to changes on feeding strategy and impaired growth. Regression also allowed a prediction of productivity loss due to percentage of birds showing scores 4 and 5, which would be around -358.6 kg per aviary, based on recently published prevalence of birds with gait scores 4 and 5. Besides the pain and discomfort experienced by the animals, severe lameness also seems to cause relevant negative impacts on production efficiency. Thus, our work confirms that severe lameness is strongly related to body weight and should be given higher priority on future husbandry and artificial selection strategies.

Commercial free-range systems for laying hens: some characteristics of nests could lead to behavioural disruptions

Céline Cayez, Vanessa Guesdon, Hélène Leruste and Joop Lensink
Groupe ISA Lille, Agriculture, 48 boulevard Vauban, 59046 Lille, France;
celine.cayez@etudiant.isa-lille.fr

Nest conditions potentially influence egg-laying sequences in poultry. Inability to express normal pre-laying behaviour could trigger disruption of this behaviour and/or abnormal behaviours such as pecking towards conspecifics. This study investigated the behaviour of laying hens inside two kinds of nests: individual nests (IN, manual egg collecting) supplied with straw or collective nests (CN, automatic egg collecting) supplied with Astroturf. We investigated 4 free-range farms (2 for each system). Babcock brown hens (n_{IN}=15 and n_{CN}=14) were video-recorded between 6.30 am and 3.30 pm in the nests. Each hen was focal-sampled from entrance into the nest to departure (egg laying sequence). This sequence was divided in two phases: 1/ pre-laying phase characterized by nest inspection, nest building (nesting substrate manipulation and scratching), 2/ laying phase characterized by laying position, egg manipulation, brooding, covering the egg with substrate. During each egg laying sequence, occurrence of pecking towards a conspecific was observed and classified as light, severe, with or without feather pulling. Mann-Whitney tests were used for data analysis. No differences were found between IN and CN hens in duration of pre-laying phase (median [Q25/Q75] = 1,280 seconds [790/2137] vs. 1,544 [921/1812]), and duration of laying phase (257 [175/405] vs. 367 [205/598]). No CN hens performed all items of pre-laying and laying phases, against 27% in IN hens. IN hens spent significantly more time than CN hens in scratching the nesting substrate (P<0.001). No IN hens performed severe pecks with feather pulling against 36% in CN hens. CN hens seemed to perform nest building behaviour by manipulating the removed feathers, in a similar way to IN hens that manipulated straw. Our results suggest that the absence of loose substrate in the collective egg-laying area could be at the origin of redirected behaviour as severe pecking with feather pulling.

Can broiler breeders on non-daily feeding schedules predict feed days?

Stephanie Torrey[1,2], Brittany Lostracco[2] and Ashleigh Arnone[1]
[1]Agriculture and Agri-Food Canada, 93 Stone Road West, Guelph, ON N1G2W1, Canada,
[2]University of Guelph, CCSAW, ANNU 249, Guelph, ON, N1G 2W1, Canada; storrey@uoguelph.ca

In North America, the majority of broiler breeders are managed with the use of non-daily feeding schedules where birds receive a large allotment of feed on some days, and no feed on other days, because of the purported success of these regimens in improving flock uniformity. These practices are banned in some countries because of the perceived negative welfare implications, yet very little is known about the welfare of breeders on different feeding schedules. The objective of this study was to compare the behaviour of broiler breeders fed on daily, '4/3' (4 days on feed, 3 non-consecutive days off feed each week) or '5/2' (5 days on feed, 2 non-consecutive days off feed) feeding schedules. In groups of 5, 75 Ross 708 broiler breeder pullets were reared from 1 d until 12 wk of age. Beginning at 3 wk, pullets were fed the same restricted amount of feed per week, but at 1 of 3 feeding frequencies. Video cameras were used to record behaviour prior to feed delivery for on- and off-feed days at 5, 7 and 11 wk. The numbers of birds drinking, foraging, walking and inactive were determined by scan sampling every 30 s. Data were analyzed with a mixed model ANOVA with week as a repeated measure. There was no effect of treatment on foraging, drinking or walking prior to feeding on on-feed days. Birds from all three feeding schedules decreased the time spent foraging and drinking, and increased the time spent walking, as they matured. There was an interaction (P=0.014) between feed frequency and time on inactivity prior to feed delivery. Those on non-daily schedules were less active over time, but daily fed birds did not change their inactivity with time, and were more active than 4/3 birds in wk 11 (P=0.043). On days when birds in non-daily treatments were not fed, there was an effect of treatment on time spent walking prior to feed delivery (P<0.0001), with daily-fed birds walking more than either of the non-daily treatments. There were interactions between treatment and time on foraging (P=0.002), drinking (P=0.0007), and inactivity (P=0.0004). While daily and 5/2 birds decreased the time spent foraging over time, those on 4/3 schedule spent the most time foraging in wk 7. All birds decreased the time spent drinking on days when non-daily treatments were not fed, but the difference was greatest for 4/3 birds who spent 58.0±4.7% of the time drinking in wk 5, and 0.8±4.7% of the time drinking in wk 11. Inactivity increased for non-daily fed birds, but not daily fed birds, and the difference was most apparent in wk 11 (P<0.01). Birds fed on non-daily schedules behaved similarly to daily-fed birds on feed days, but appeared to anticipate non-feeding days.

The association between feather-eating and severe feather-pecking in free-range laying hens

Kate Hartcher[1], Stuart Wilkinson[1], Paul Hemsworth[2] and Greg Cronin[1]
[1]*University of Sydney, Faculty of Veterinary Science, 425 Werombi Road, Camden, NSW 2570, Australia,* [2]*University of Melbourne, Animal Welfare Science Centre, Alice Hoy Building, Parkville, VIC 3010, Australia; kate.hartcher@sydney.edu.au*

Severe feather-pecking (SFP) is a detrimental behaviour whereby birds peck at and pull out the feathers of other birds. Feathers are often ingested once they have been removed. Severe feather-pecking has been identified as the biggest welfare concern for laying hens, and the most prevalent behavioural problem in the laying industry. As SFP is thought to be multi-factorial, potential causative factors should be investigated. Although it has been suggested that feather-eating (FE) is the primary motivation for SFP, few studies have investigated the relationship between SFP and FE. This experiment investigated the level of interest in loose feathers by SFP birds compared to non-SFP birds in two free-range flocks of ISA Brown laying hens. Six pens of 50 hens were tested in each flock at approximately 42 weeks of age. Individual feathers were introduced to the home pens and placed in the centre of each pen. This was repeated 15 times per pen in a random order. Data were analysed in Genstat using logistic generalized linear mixed models for binomial data, and in R with survival analysis using Cox's proportional hazard test for latencies. The SFP birds in flock one had shorter mean latencies to peck at (P<0.001, 5 vs 15 s) and ingest (P=0.001, 19 vs 25 s) feathers. Similar results were found in flock two, where SFP birds had shorter latencies to peck at (P<0.001, 3 vs 21 s) and ingest (P<0.001, 6 vs 23 s) loose feathers. In flock two there was also a trend for SFP birds to consume feathers more frequently than non-SFP birds (P=0.085, 97% vs 24%). The results of this experiment support previous studies that FE is positively correlated with SFP and may aid diet formulation and inclusion of forages to prevent SFP.

Behavioural and physiological responses to stress during different life phases in chickens (*Gallus gallus*)

Maria Ericsson[1], Rie Henriksen[1], Ann-Sofie Sundman[1], Kiseko Shionoya[2] and Per Jensen[1]
[1]IFM Biology, AVIAN Behavioural Genomics and Physiology group, Linköping University, 581 83 Linköping, Sweden, [2]Department of Clinical and Experimental Medicine, Division of Cell Biology, Linköping University, 581 83 Linköping, Sweden; miaer@ifm.liu.se

Stress early in life disrupts normal brain and HPA-axis development and can cause behavioural alterations and changes in HPA-axis reactivity. In chickens, unstable environments early in life are known to cause life-long phenotypic alterations. During the transition to sexual maturity, a cascade of developmental processes occurs, indicating a potentially sensitive period. Increasing evidence also point towards transgenerational effects of early stress. In chicken farming, potential stressors are abundant throughout life but the consequences of stress during adolescence are not known. We therefore aimed to compare short- and long-term effects of stress applied at different ages. Hy-Line chicks (n=200), group housed in one floor pen, were assigned to three treatment groups and one control group (C). The treatments consisted of exposure to one week of stress treatments (restraint, social isolation, food frustration) either during early development (2W), early adolescence (8W) or late adolescence (16W). From 21 weeks of age, all birds were tested for behavioural alterations and corticosterone response. An F1-generation was hatched to investigate transgenerational effects. Immediate reactions to stress (one week after treatment) included reduced growth in 2W (P<0,001) and 8W (P<0,01) and decreased fearfulness (2W, P<0,01; 8W, P<0,05) compared to C when analyzed with an ANOVA. Also, 8W were more active (P<0,05). As adults, 8W had a stronger corticosterone response to physical restraint than the other groups (P=0,01). Transgenerational effects in HPA-axis reactivity were seen in the 8W group (P<0,05). In conclusion, immediate behavioural effects were strongest in the early adolescence group (8W), and persistent effects were seen in HPA-axis reactivity in these birds as adult as well as in their offspring. We therefore conclude that adolescence is a sensitive period in chickens, since stress exposure induces both short- and long-time phenotypic shifts.

Dust-bathing behavior and feather lipid levels of laying hens in enriched cages

Richard A Blatchford, Margaret A De Luz and Joy A Mench

University of California, Davis, Center for Animal Welfare, Department of Animal Science, Davis, CA 95616, USA; jamench@ucdavis.edu

Laying hens in enriched colony cages (EC) are often provided with a feed-sprinkled Astroturf pad to encourage foraging and dust-bathing. As part of the Coalition for a Sustainable Egg Supply project, we evaluated the use of this pad for dust-bathing and its effectiveness in reducing feather lipid levels in 60-hen EC on a commercial farm. We also assessed whether exposure to litter during rearing affected dust-bathing or feather lipids in EC. Breast and back feathers were cut from 19-wk-old pullets that had been raised in either an aviary (AVP) or a conventional cage (CCP) pullet-rearing facility (n=27 pullets/system). The pullets from CCP were then placed in conventional (CC) or EC (n=6 cages) on the laying hen farm, while those from AVP were placed in an aviary (AV) or EC (n=6 cages) on the same farm. The 12 EC were then observed during the first 3 days after the pullets were placed, as well as when the hens were 72 wk of age, to determine the number of birds dust-bathing on the pad. Instantaneous scan samples were made at 1-min intervals for 30 minutes/cage from 08:30-10:30, 12:45-14:45, and 16:00-18:00. Feather lipid levels were evaluated in the same EC when the hens were 72 wk of age, as well as in hens from the AV and CC (n=27 hens/system), using Soxtec extraction. Data were analyzed using ANOVA and t-tests. Surprisingly, pullet rearing conditions had no effect on feather lipids at placement, with AVP and CCP having similar lipid levels on their breasts (AVP: 6.68±0.25, CCP: 7.34±0.45 mg lipid/g feather, P=0.21) and backs (AVP: 6.00±0.31, CCP: 5.71±0.21 mg lipid/g feather, P=0.45), even though AVP had been raised with wood-shavings litter. Rearing system also did not affect the number of hens using the pad for dust-bathing either immediately after placement (pooled mean: 0.91±0.06 hens/30 min, P=0.84) or at 72 wk of age (pooled mean: 0.79±0.12 hens/30 min, P=0.61). At 72 wk, breast feather lipid content was lower (P=0.003) for AV hens (16.5±0.5 mg lipid/g feather) than CC hens (21.3±1.8 mg lipid/g feather), with EC (18.6±0.9 mg lipid/g feather) hens intermediate. Back feather lipids were lower (P<0.001) for AV hens (13.2±0.5 mg lipid/g feather) than for both CC (21.3±1.8 mg lipid/g feather) and EC (16.6±0.6 mg lipid/g feather) hens. Provision of the Astroturf pad therefore resulted in some reduction of lipid levels on the breast feathers despite the fact that it was not extensively used for dust-bathing, but back feather lipid levels in EC hens were similar to those of CC hens not provided with dust-bathing material. These results suggest that Astroturf with feed is minimally used and only partially effective as a dust-bathing substrate in EC.

The effects of a mother-hen on behavior and welfare of chicks

Patrick Birkl
University of Innsbruck, Ecology, Technikerstraße 15, 6020 Innsbruck, Austria;
patrick.birkl@student.uibk.ac.at

Social interactions between broody hens and their chicks are very complex and have been the subject of numerous studies. This experiment investigated the behavioural and physiological effects of the presence or absence of a broody hen on chicks after their third week of life. Five batches of eggs from Australorp chickens were brooded and assorted to either a broody Brahma hen (Group1$_{n=10}$, 2$_{n=9}$ Degrees of freedom are fractional to adjust for violations of sphericity assumptions and 3$_{n=10}$) on day 17 of brood or remained in the brooder to hatch (Group 4$_{n=7}$ and 5$_{n=11}$). Behaviour and vocalization was video-recorded for at least 5 hours continuously on day 19, 28, 41, 51, 64 and 74(1 minute=1 observation; 2×120 min per day). Corresponding ethograms were created in order to analyze and compare active- foraging- and resting-behaviour. Vocalizations were analyzed using 'Sonogram Visible Speech 3.0' software. Stress related vocalizations were then recorded for each group (2×2 hours per day). Furthermore, blood-smears were used to determine Heterophile to Lymphocyte Ratios (H/L) of each chick (Group 3, 4 and 5) at the end of the experiment. Total active foraging behaviour of motherless chicks was significantly lower than in brooded chicks (mean ±SD 0.258±0.113, 0.41±0.08, t=1200 min, P<0.005). Also, motherless chicks spent significantly more time resting (mean ±SD0.435±0.156, 0.355±0.081, t=1200 min, P<0.005) than brooded chicks. Furthermore brooded chicks spent more time feeding in presence of the hen e.g. [Day 41: Group 3 (mean ±SD 0.238±0.216) versus Group 5 (mean ±SD 0.189±0.130), and less time after the hen was removed [Day 74: Group 2 (mean ±SD 0.211±0.17) versus Group 5(mean ±SD 0.129±0.16)]. Blood-smear analysis showed no significant differences between motherless and brooded chicks; Group 3 (mean ±SD 0,516±0,298), Group 4 (mean ±SD 0,678±0,367), Group 5 (mean ±SD 1,096±0,602), probably due to small sample size (n$_{total}$=28). Occurrence of high-frequency distress calls appeared to be insignificant between brooded and motherless chicks. The present study demonstrated the critical role of the mother hen in motivating foraging behaviour of chicks. This result offers new avenues for mother-chick interaction research.

The effects of unhusked rice feeding on behavior and quality of feces in broilers

Ai Ohara[1], Miho Nishi[2], Satsuki Tomita[2], Seiji Nobuoka[2] and Shusuke Sato[1]
[1]Tohoku University, Graduate School of Agricultural Scince, 232-3 Naruko-onsen, Yomogita, Osaki, Miyagi, 989-6711, Japan, [2]Tokyo University of Agriculture, 1737 Funako, Atsugi, Kanagawa, 243-0034, Japan; oharala@gmai.com

We found that FPD might be few in broilers fed the unhusked rice. In this study, we tested whether feeding the unhusked rice stimulates behavior and improves qualities of feces in broiler chickens. 10 birds in single cages were allocated equally to 2 dietary treatments: Control (general feed: 60% of feed was the corn) and UHR (half of the corn was replaced by the unhusked rice). We recorded behavior for 2 h each after feedings at 09:00 h and 15:00 h at 4- and 7-weeks old. Birds were scanned at 1 min intervals on maintenance behavior. No of meals, pecking and behavior sequence (BS) with feeding behavior were recorded continuously for 30 min after each feeding. Moisture and nitrogen concentrations in feces, amount of feed consumed and H/L ratio were measured. At 60 days old, each chicken was provided a novel object on its trough and observed duration of looking at it continuously for 3 min. Mann-Whitney U test was mainly used for analysis. There was no significant difference between treatments at 4 weeks old. At 7 weeks old, preening (%) tended to be longer (control 1.0±0.7, UHR 2.0±1.3; multiple logistic regression, P=0.052), No of meal was lower (292±122.9, 173.2±86.4; t-test, P<0.05) in UHR than Control. No of pecking was not different between treatments. BS from feeding to preening (No) was significantly fewer in UHR (cell-by-cell test, P<0.05). Total duration (s) of looking a novel object was shorter (8.6±2.6, 3.8±2.9; P<0.05) and H/L ratio was lower (0.35±0.11, 0.14±0.08; P<0.05) in UHR than control. Moisture and nitrogen of feces (%) were lower (76.9±4.5, 5.0±0.3) in UHR than Control (81.6±2.7, 6.4±0.8)(P<0.05, P<0.01). Amounts of feed consumed were not different between treatments.

Applying animal welfare science to ethical debate: the animal protection approach

Mark Kennedy

World Society for the Protection of Animals, 222 Gray's Inn Road, WC1X 8HB London, United Kingdom; markkennedy@wspa-international.org

Animal welfare science allows us to objectively assess the welfare of animals. In isolation it informs, but does not resolve, ethical debate on which levels of welfare are morally acceptable. Utilitarian animal welfare philosophy holds that it is acceptable for humans to use animals if the use in question is justified by cost-benefit analysis, and asserts that animals must be treated as humanely as possible. A limitation is the difficulty of reconciliation of utilitarian pleasure and pain currencies between beneficiaries and victims. Furthermore, the utilitarian approach is heavily influenced by subjective evaluation of the relative value of animal lives. It might seem that such troublesome uncertainties are avoided in the deontological animal rights approach. In holding that certain (or all) uses of animals are wrong in principle, questions of the amount of suffering become irrelevant. If animals should not be used in such ways, any suffering is unnecessary and any benefit irrelevant. Organisations with a rights-based approach to animal use often encounter closed ranks when attempting to influence animal users, limiting their contribution to the pragmatic improvement of the lives of animals. As an alternative approach, it could be argued that the case for our having moral obligations to animals rests in their sentience and telos. What we do to them and how we make them live matters because they have positive and negative experiences and emotions. They have telos- nature (which includes species-typical behaviour) and interests- as well as intrinsic value beyond their utility to humans. At the very least animals have an interest in having a 'Good Life'. The role of animal protection organisations such as WSPA is to work to promote and protect animal interests, which are often at risk of being over-ridden by competing interests.

Persistent effect of broody hens on behaviour of chickens

Toshio Tanaka[1], Tsuyoshi Shimmura[1,2,3,4], Yuji Maruyama[1], Saori Fujino[1], Eriko Kamimura[1] and Katsuji Uetake[1]
[1]*Azabu University, Animal Science and Biothechnology, 1-17-71 Fuchinobe, Chuo-ku, Sagamihara-shi, 252-5201, Japan,* [2]*National Institute for Basic Biology, Division of Seasonal Biology, Nishigonaka, Myoudaiji-cho, Okazaki-shi, 444-8585, Japan,* [3] *The Graduate University for Advanced Studies, Hayama-cho, Miura-shi, 240-0193, Japan,* [4]*Nagoya University, Graduate School of Bioagricultural Science, Furo-cho, Chikusa-ku, Nagoya-shi, 464-8601, Japan; tanakat@azabu-u.ac.jp*

We reported previously the details of dynamic changes of behaviour of brooded and non-brooded chicks at an early age and indicated that behavioural development of chicks was promoted remarkably by the presence of a broody hen. Here we reported that these effects at an early age persist after maturity. A total of 60 artificially hatched female chicks were randomly assigned to one of two treatment groups: six pens with five chicks (brooded group) each were reared by a broody hen and six pens with five chicks (non-brooded group) each were provided with an infrared heating lamp. We evaluated the enduring effects of broody hens by measures of behaviour, physical condition and production at 9, 16, 35, and 55 weeks of age. The numbers of threatening, aggressive pecking, fighting, and severe feather pecking behaviours were higher in non-brooded than in brooded group (all P<0.05). The numbers of aggressive pecking and severe feather pecking incidents tended to increase with age and the number of severe feather pecking incidents was higher at 35 and 55 weeks than at 9 and 16 weeks of age (all P<0.05). The egg production was lower in brooded than in non-brooded chickens (P<0.05), while the number of brooding chickens was higher in the brooded than in the non-brooded group (P<0.05). In conclusion, the presence of broody hens at early stage of chicks' live has the enduring effect on behaviour. Although brooded chicks showed more brooding and egg production was decreased, feather pecking and aggressive interaction were decreased. On the basis of these results, the development of more plentiful simple enrichment methods for artificially hatched chicks that modify behaviour but do not decrease egg production are needed in the future.

The physiological and behavioral responses to the administration of cisplatin, an anticancer drug that induces nausea

Masato Aoyama[1], Minami Shioya[1], Atsushi Tohei[2] and Shoei Sugita[1]
[1]Utsunomiya University, Utsunomiya-shi, Tochigi, 321-8505, Japan, [2]Nippon Veterinary and Life Science University, Musashino-shi, Tokyo, 180-8602, Japan; aoyamam@cc.utsunomiya-u.ac.jp

Nausea is a common indicator of diseases and disorders. Ruminants, which play an important role as livestock, have not been thought to feel nausea, because they might not have a vomiting reflex. The aim of this study is finding some physiological and/or behavioral response that represent nausea in ruminants. Five adult Shiba goats (two males and three females) were used. Each animal was given cis-diammineplatinum (II) dichloride (cisplatin: cisP), which is one of the cancer chemotherapy drugs but induce nausea in humans, dogs and ferrets, intravenously (1 mg / 0.5 ml saline / kg BW). The same volume of saline was administered for the control, and at least 10 days separated control and cisP sessions. The behavior of goats were videotaped and analyzed by continuous observation. Blood samples were corrected for the levels of the cortisol (Cor) which is a physiological marker of stress. Because previous reports indicate that peptide YY (PYY), cholecystokinin (CCK) and arginine vasopressin (AVP) are involved in the induction of nausea and vomiting in humans and dogs, we also measured the blood levels of these peptides. Goats administrated with cisP did not show any remarkable behavioral changes immediately after the administration, but about 100 min after, all goats assumed a specific posture consisting of a lowered head and very little movement, even when we approached them for the blood collection. Blood level of Cor did not change 60 min after the administration,. but it significantly increased after 120 min. These changes in behavior and blood Cor level were not observed in control goats. However, cisP administration failed to affect the blood levels of PYY, CCK and AVP throughout the 240 min experimental period. These results indicate that goats seem to feel nausea by administration of cisP like other animals, and the body posture can be one of the indicator to evaluate nausea in goats. However, none of blood level of PYY, CCK nor AVP can be the physiological marker to represent the nausea in goats.

Effect of body condition and gait score on estrous expression parameters measured by an automated estrous detection system

Tracy A. Burnett, Augusto M.L. Madureira, Bruna F. Silper, Artur C. Fernandes and Ronaldo L.A. Cerri
University of British Columbia, Faculty of Land and Food Systems, 2357 Main Mall, V6T1Z4, Canada; tracyanneburnett@gmail.com

The aim of this study was to determine if body condition, gait and hock scores in early lactation of dairy cows affected the expression of estrous events detected by an automated estrous detection system (Heatime, SCR Engineers, Israel). Three hundred and twenty-three lactating Holstein cows were equipped with collar-mounted automated activity monitors. The activity monitor was equipped with an accelerometer that continuously recorded physical activity and detection of estrus episodes (n=423) and data submitted to the central station every two hours by wireless communication. Animals had their estrus synchronized with two injections of PGF2a and animals had their body condition score (BCS; scale 1-5), hock score (HS; scale: 1-4), gait score (GS; scale:1-4) and the presence of corpus luteum by ovarian ultrasonography recorded. Estrous expression was quantified using two parameters: (1) peak activity; and (2) duration of the estrous episode. Peak activity was defined as the maximum activity index during an estrous episode; the threshold activity to be considered an estrous event was set at an index level of 35, approximately 80% increase in activity compared with the baseline. The duration of an estrous episode was defined as the amount of time the animal spent with an index level greater than 35. Data was analyzed using ANOVA for repeated measures and logistic regression of SAS. Contrary to our initial hypothesis, animals considered lame (GS>2) and those with low amounts of body fat reserve (BCS<3) did not differ on the duration and peak intensity of estrus episodes (P>0.10). Similarly, estrous duration and intensity was similar between cyclic and non-cyclic cows by 50 days postpartum (P>0.40). The proportion of animals above or below the medium intensity (index=80), and duration (episodes=10 h) was also unaffected by lameness, BCS and cyclicity. Cows that became pregnant from the first postpartum AI had a greater average peak intensity of their estrus episodes previous to AI (74.5±1.8 vs 70.5±1.5; P=0.03). Average duration was not affected by pregnancy success outcome. Pregnancy per AI at first breeding postpartum was decreased in animals with BCS<2.75 (46.2% vs 28.7%; P<0.01), gait score>2 (40.4% vs 26.1%; P=0.03) and non-cyclic by 50 days postpartum (39.4% vs 15.4%; P<0.01). In conclusion, BCS, gait score and cyclicity strongly affected pregnancy per AI, however, intensity and duration of estrus episodes measured by an activity monitor system were not affected by these parameters.

Resting pattern of nulliparous dairy cows changes with estrus

Bruna Silper, Liam Polsky, John Luu, Anne Marie De Passillé, Jeffrey Rushen and Ronaldo Cerri
University of British Columbia, Faculty of Land and Food Systems, 2357 Main Mall, V6T 1Z4
Vancouver, BC, Canada; brunasilper@gmail.com

Technological developments allow cattle activity to be automatically monitored, increasing the potential for automated estrus detection. Lying and standing behaviour of nulliparous Holstein dairy cows (n=30) were studied around the time of estrus (135 estrus episodes) to determine patterns of overall activity during estrus using an accelerometer attached to the rear leg (IceTag, IceRobotics, Scotland, UK). Heifers were kept in a free stall barn in groups of 7 to 12, from 6 to 12 months old. Frequency and duration of lying and standing bouts were analyzed. Data exported from accelerometers was used to calculate, on a 24 h basis, the mean, median and standard deviation of bout duration, duration of the longest bout for both standing and lying, and total time spent in each position. Results are presented as lsmeans ± SE for days -7 (baseline), -2, -1, 0 (estrus), +1, and +2 relative to estrus. Estrus periods were previously identified from activity peaks (increased number of steps) and validated with biweekly ovarian ultrasonography. Data was analyzed by ANOVA for repeated measures using the mixed procedure of SAS. Frequency of lying and standing bouts was smaller on estrus day compared with other days ($P<0.01$). Bout frequency started to reduce at day -1 (9.8 ± 0.4 bouts/d), since bout frequency at baseline and estrus was 11.3 ± 0.4 and 8.1 ± 0.4 bouts/d, respectively ($P<0.01$). There was no effect of estrus on average duration of lying bouts ($P>0.10$). However, average duration of standing bouts increased from 56 ± 3 min during baseline to 76 ± 3 min at day -1 and 127 ± 3 min at day 0 ($P<0.01$). Mean duration of standing and lying bouts on days +1 and +2 was not different from baseline ($P>0.10$). Standard deviation for standing bout duration was greater for days -1 and 0 than for other days ($P<0.01$). This variable was also greater during estrus than at day -1 ($P<0.01$), meaning that bout duration is more variable during estrus than during non-estrus days. Total lying time reduced from 845 ± 13 min during baseline to 577 ± 13 min during estrus. The reduction began before the estrus day, with 756 ± 13 min of lying on day -1 ($P<0.01$). Duration of the day's longest standing bout seemed to be the variable most influenced by estrus (489 ± 12 min on day 0 vs. 251 ± 12 min during baseline; $P<0.01$). The longest bout represented on average 52% of the total standing time compared with 35% in non-estrus periods. Lying and standing behaviour of heifers is markedly influenced by estrus. Behavioural changes were observed during late proestrus, suggesting that automated measures of cow activity may result in earlier and more precise estrus detection.

Effect of three different housing systems on dairy cattle welfare

Anna Fernández, Eva Mainau, Xavier Manteca, Adriana Siurana and Lorena Castillejos
Universitat Autònoma de Barcelona, Department of Animal and Food Science, Campus UAB.
Edifici V. V0-135, 08193 Bellaterra (Cerdanyola del Vallès), Spain; anna.fernandez@uab.cat

Dairy housing systems have an important impact on cow comfort. The objective of this study was to compare cow comfort regarding the effects of three different housing systems for lactating cows: compost bedded pack (CB), traditional bedded pack (TB) and cubicle (C). The study was conducted on a commercial farm with 814 lactating cows where 2 barns for each treatment were evaluated. Data was obtained in summer and autumn during one day each period by four previously trained observers. Behaviour indicator (time needed to lie down) and health indicators (dirtiness of the cows, body condition, integument alterations, lameness, respiratory, digestive and reproductive disorders) based on Welfare Quality® assessment protocol for cattle were evaluated. Statistical analyses were carried out with the SAS software using a GLIMMIX procedure. Cows spent similar time to lie down irrespective of the treatment and the season studied (4.7 ± 0.09 seconds; $P>0.05$). Comparing systems, the percentage of cows with dirtiness on hindquarters and the udder was lower ($P<0.001$) in C ($19.7\pm4.35\%$ and $25.0\pm4.37\%$) than in CB ($54.5\pm5.72\%$ and $44.1\pm4.43\%$) and TB ($50.4\pm7.73\%$ and $56.4\pm6.35\%$). Cows allocated in C showed the highest prevalence of hairless patches on the tarsus ($55.4\pm6.41\%$; $P<0.01$) and lesions or swellings throughout the body ($51.4\pm5.15\%$; $P<0.01$). Comparing seasons, percentages of cows with dirtiness and with hairless patches on the flank were higher in autumn than in summer ($P<0.05$). In summary, animals in C were cleaner, but have more tegument alterations than animals in CB and TB, making it difficult to conclude if there is any overall difference in animal welfare across systems and suggesting that more studies are needed in this area.

Provision of nest-building materials and space prior to parturition could reduce the number of stillborn piglets?

Jinhyeon Yun, Kirsi Swan, Olli Peltoniemi, Claudio Oliviero and Anna Valros
University of Helsinki, Department of Production Animal Medicine, P.O.Box 57, 00014, University
of Helsinki, Finland; jinhyeon.yun@helsinki.fi

Several studies reported that an increase of sow plasma oxytocin concentrations could alleviate anxiety during parturition, and also reduce farrowing duration. We investigated whether provision of nest-building materials and space prior to parturition would increase sow plasma oxytocin concentrations during prepartum and parturition, and whether elevated circulating oxytocin concentrations could reduce the number of stillborn piglets. We allocated 33 sows in (1) CRATE: the farrowing crate closed (210×80 cm), with provision of a bucketful of sawdust, (2) PEN: the farrowing crate opened, with provision of a bucketful of sawdust, (3) NEST: the farrowing crate opened, with provision of abundant nest-building materials. We collected sow blood samples for an oxytocin assay via indwelling ear vein catheters on days -3, -2, -1 from parturition twice a day, and at 0, 2, and 4 min after the first five piglet birth during parturition. Litter size, still born, and born alive ratio were recorded. Oxytocin concentrations were analysed with a mixed model using repeated measures. Pearson correlation coefficients (r) were used to examine interactions of oxytocin concentrations between prepartum and parturition. Litter performance was tested to determine interactions with oxytocin concentrations by Spearman rank correlation coefficients (r_s). Sow plasma oxytocin concentrations in NEST were significantly greater than in CRATE during the prepartum period (P<0.05). Prepartum oxytocin concentrations were significantly correlated with farrowing oxytocin concentrations (r=0.64, P<0.0001). However, oxytocin concentrations during the first five piglets born did not differ among three treatments (P>0.10). Litter performance was not significantly different among three different farrowing environments (P>0.10), and had no correlations with sow prepartum or parturition oxytocin concentrations. These results indicate that provision of nest-building opportunity prior to parturition could increase prepartum sow plasma oxytocin concentration, possibly through allowance of natural behaviour, but did not affect to oxytocin concentration during parturition and litter performance.

The effects of food type on wool-biting in housed sheep

Chen-Yu Huang and Ken-ichi Takeda
Shinshu University, graduate school of agriculture, Minamiminowa-mura, 8304, Kamiina-gun,
Nagano, Japan, 399-4598, Japan; 12st553a@shinshu-u.ac.jp

Although wool-biting is a serious animal welfare problem in housed sheep, the factors that induce this abnormal behaviour are still unknown. As inappropriate feed may lead to some abnormal behaviours in other domestic animals, in this study, we investigated the effect of food type on wool-biting in housed sheep. A flock of 10 Friesland ewes raised in a 24 m^2 pen bedded with 10 cm deep of straw was taken. Three food types were used as treatments as follows: compact hay bales cutted in 30 cm, hay rolls, and hay cubes. Each animal received their treatment feed for 400 g at 08:30 and 600 g at 15:00, and 350 g of concentrate after afternoon meal every day. Each treatment was fed for four weeks, followed by 1 week for adaptation to the next treatment. During the adaptive week, the straw bedding was replaced, except that during treatment 3 it was refilled every week for sanitation because the sheep ate all of the bedding. Behavioural observations were taken on two days each week at 06:30-08:30, 09:30-11:30, 13:00-15:00, and 16:00-18:00. The numbers of biting bouts, and biting numbers within a bout were recorded. The biting bouts per hour and the total biting numbers per hour in treatment 1 were significantly greater than in treatments 2 and 3 (repeated-measures ANOVA: P<0.01; treatment 1: 22.0 and 57.7; treatment 2: 4.3 and 9.9; treatment 3: 6.9 and 13.6). In treatment 2, sheep fed on hay rolls showed lower frequency of wool-biting post-feeding (paired t-test: P<0.01). It may be the interlocking hay tangled with each other in the form of roll thus provide sufficient mouth stimulation for sheep. Although sheep fed on hay cubes in treatment 3 showed a low frequency of wool-biting, they transferred their behaviour to bed-eating which was never seen in treatments 1 and 2, and perhaps because of the oral stimulation re-directed to the straw bedding, resulting in the low tendency to pull wool from other individuals. Our results suggest that although it is difficult to stop wool-biting in a flock that is already habituated to it, appropriate feed can repress this behaviour, and that hay rolls may be an effective food type for housed sheep.

Corn addition during last 15 days of gestation in the diet of Romanov ewes, improves the early maternal selective behaviour

Ethel García Y González[1], Angélica Terrazas[2], Edith Nandayapa[2], Alfredo Cuellar[2], Said Cadena[2], Rita Bonelli[2], Jorge Tórtora[2], J. Alberto Delgadillo[1], J. Alfredo Flores[1], Jesús Vielma[1], Gerardo Duarte[1] and Horacio Hernández[1]
[1]UAAAN, CIRCA, Periférico Raúl López Sánchez, 27054, Torreón, Coahuila, Mexico, [2]FESC,UNAM, Posgrado en Ciencias de la Producción y de la Salud Animal, Km 2.5 Cuautitlán-Teoloyucan, 54714, Edo. de México, Mexico; eth_cat@hotmail.com

Maternal selective behaviour in ewes refers to the fact that mothers accept only their own lamb at the udder and actively reject aliens. This phenomenon has been widely studied in breeds of ewes non-considered as prolific breeds. Therefore, we investigated if in nulliparous Romanov ewes, a prolific breed, a corn addition (0.5 k/ewe/day) in the diet during the last 15 days of pregnancy affects maternal selective behaviour at 4 h post-partum. Mothers from the control group (CG; n=7) and from the corn supplemented group (SG; n=8) were used. Every mother was put in the test pen (2 m^2) and the selectivity test consisted of 2 successive presentations to own lamb or an alien for 3 minutes. The behaviours recorded in the mothers were: low and high pitch-bleats, time allowed to the lamb to be near the udder, frequency of rejections and acceptances to the udder and aggressive behaviours. When comparisons were made between groups, there was not any statistical difference in these variables (P≥0.15). However, when comparisons were made into each group, we found that the mothers of the CG did not show difference between the number of low pitch-bleats in the presence of their own lamb or when was exposed to an alien one (29±9.2 vs 15±8.5 times; P=0.25). Mothers of the SG emit more low pitch-bleats with their own lamb than with the alien (20±4.1 vs 5.3±3.1 times; P=0.05). In both groups, mothers emit more high pitch-bleats in presence of the alien lamb than with their own lamb (CG: 40±8.8 vs 5±1.6 times; SG: 39.0±6 vs 5.6±2.5 times; P≤0.004). In both groups, the time that mothers allow their own lamb stay near the udder was higher than to the alien (CG: 58±19 vs 3.5±2.3 s; SG: 96±18.0 vs 2±1.8 s; P≤0.03). In mothers from SG, the frequency of the rejections to the udder was higher for the alien lamb than to their own lamb (2±0.6 vs 0±0 times; P=0.01), while in mothers from CG the same values did not reach significance (P=0.08). The mothers of the SG accepted to the udder with more frequency to own lamb than the alien one (2.3±0.4 vs 0.1±0.1 times; P=0.006. Finally, mothers of the SG show more aggressive behaviours to the alien lamb than to the own (10±3.5 vs 0±0 times, P=0.001). We conclude that even in ewes from a prolific breed as Romanov, corn addition in the diet during last 15 days of gestation improves the early selective behaviour of the mothers. Grant UNAM PAPIIT IN217012.

Neglect behaviour of Japanese Black cattle (*Bos taurus*): features and status

Daisuke Kohari[1], Azusa Takakura[1] and Ken-ichi Yayou[2]
[1]Ibaraki University, Inashiki-gun, Ami-machi, 3000331, Japan, [2]National Institute of Agrobiological Science, Tsukuba, 3058602, Japan; dkohari923@gmail.com

Neglect of young is an important problem for animals in captivity. Nevertheless, the causes and diversity of such behaviours remain unclear. We investigated those points using a questionnaire survey administered to beef cattle owners. We surveyed the breeding experience of neglecting cattle at 703 cattle breeding farms in Ibaraki prefecture, Japan. Subsequently, we interviewed 31 cattle workers of the first survey to assess appearance states and behavioural patterns of neglect. Furthermore, the owner's or worker's impressions of neglectful cattle were elicited and then evaluated using 11 pairs of adjective phrases related to the features of their neglectful cattle with a five-point scale. Impression data were assessed using semantic differential (SD) method. Factor analysis was conducted to extract features of neglectful cattle based on SD data (Package 'psych', R freeware, ver. 3.0.2.). Of 703 subject farms, 289 responded to the first survey (41.1% response), 69 of which reported keeping neglectful cattle (24%). Four appearance states of neglect were reported: only in primiparous animals (22%), unexpected (6%), chronic (66%), and improved (6%). Behavioural patterns of neglect were also divided into four types: aggression / no grooming (75%), no aggression / no grooming (10%), aggression / grooming (6%), and no aggression / grooming (9%). Results from factor analysis of SD method extracted three factors. Factor 1 was constructed using seven temperament-related phrase pairs such as 'agitated–cool', 'active–inactive', and 'gentle–fierce'. It was interpreted as 'Gentle' based on results of semantic profiles. Factor 2 was constructed using evaluative related phrase pairs such as 'cowardly–brave' and was interpreted as 'Friendliness'. Factor 3 was constructed using three activity related phrase pairs, such as 'curious–indifferent' and 'habituated – not habituated', and was interpreted as 'Activeness'. These contribution rates were 24% for factor 1, 17% for factor 2, and 13% for factor 3. These results demonstrate that neglect among Japanese Black cattle exists with high frequency. Their appearance and behavioural patterns were not uniform. Three impressive factors were inferred as features of neglecting cattle.

Effects of prenatal handling stress on maternal provisioning in dairy goats

Emma M Baxter[1], Adroaldo J Zanella[2], Jo E Donbavand[1] and Cathy M Dwyer[1]
[1]SRUC, Animal Behaviour and Welfare Team, Kings Buildings, West Mains Road, EH9 3JG Edinburgh, United Kingdom, [2]USP, Av Duque de Caxias Norte, 225, 13635-900 Pirassununga SP, Brazil; emma.baxter@sruc.ac.uk

Goats have generally been over-looked in the field of prenatal stress research, however dairy goats are subjected to a number of potential stressors throughout their lives, including daily interactions with humans. The quality of these interactions may have direct consequences for the animal undergoing the experience, but if such events occur during gestation it can adversely affect the developing fetus. Therefore this study examined the effects of differential handling during mid-gestation in 40 twin-bearing Saanen × Toggenburg primiparous goats. Between days 80-115 of pregnancy goats were subjected to a negative (N, n=13), positive (P, n=13) or minimal (M, n=14) handling protocol for 10 minutes twice a day. The control (M) group did not receive handling treatments and all goats were subjected to normal husbandry procedures only outside treatment periods. Behavioural and physiological data were collected from both the mothers and their kids, including detailed measurements of mother-offspring interactions for the first 2 hrs post-partum. Following delivery placentae were dissected and morphology analysed. The groups were unbalanced so Linear Mixed Models (REML) were used to determine the statistical differences between the treatment groups. Data were adjusted for litter size and sex ratio where necessary. Negatively handled goats lost 4 fetuses before birth compared to no fetal loss in other groups (Fishers exact P=0.051). Treatment also influenced placental morphology with a tendency for fewer medium raised cotyledons evident in the negative treatment (n=59.6 (sem±8.5) vs. P=77.0 (sem±6.4) or M=76.0 (sem±9.9); $F_{2,34}$=3.04, P=0.06). Positively-handled goats were more attentive to their kids displaying significantly more grooming behaviour towards their young during the first two hour observation period post-partum (P=92.1% (sem±6.1%) vs. n=72.5% (sem±7.0%) vs. M=61.7% (sem±8.4%); $F_{2,33}$=3.28, P=0.05). Preliminary analysis of kid behaviour suggests no treatment effects on neonatal behaviours and vocalisations. The results show that handling during pregnancy affects maternal provisioning in the form of placental quality and post-partum maternal behaviours. Such results have important animal welfare implications, demonstrating that negative handling of pregnant females results in poorer placental quality with potential for fetal loss. It also demonstrates the beneficial effects of positive handling on important maternal behaviours that have been shown in other ruminants to influence offspring survival and development.

Space allowance in gestating ewes and early postnatal separation: Impact on the coping abilities of newborn lambs

Xavier Averós[1], Joanna Marchewka[1], Ignacia Beltrán De Heredia[1], Roberto Ruiz[1] and Inma Estevez[1,2]

[1]Neiker-Tecnalia, Animal Production, Arkaute Agrifood Campus, P.O. Box 46, 01080 Vitoria-Gasteiz, Spain, [2]IKERBASQUE, Basque Foundation for Science, Alameda Urquijo 36-5 Plaza, 48011 Bilbao, Spain; xaveros@neiker.net

A common prenatal stressor in lambs may be the space allowance given to ewes during pregnancy, followed by early maternal separation. They may affect lamb development and future coping abilities. We studied the impact of different space allowances during pregnancy and early maternal separation in lambs' coping abilities. A total of 54 ewes were housed at 1, 2, or 3 m^2/ewe during the last 11 weeks of pregnancy (SA1, SA2 and SA3). After birth, lambs remained with ewes (MRL), or were transferred to 1 of 3 artificially rearing pens, after colostrum consumption, according to prenatal treatment (ARL). At about 3 days of age, 42 lambs (unbalanced across treatments) were divided into groups of 3, imposing 2 restrictions: (1) age was at least 24 h; and (2) ARL spent at least 24 h in the artificial rearing pens. Each lamb was subjected to a novel arena (2.5 min) and a novel object (2.5 min) test. Behaviour, use of space, and vocalizations were collected. After this, lambs were exposed to a social motivation test (5 min), where the arena was divided into 3 regions according to the proximity to an adjacent pen where the other 2 lambs waited. Number of visits, time spent/region, and vocalizations were collected. Generalized mixed model ANOVAs were used to test the effects of pre, postnatal treatments, and gender. Frequency of standing immobile in SA1 ARL was higher than SA1 MRL (0.81±0.08, 0.60±0.12; $P<0.05$) during novel arena tests. ARL spent more time close to other lambs during social motivation tests, though values for SA3 and SA2 ARL were lower than those of SA1 ARL (78.4±1.7, 72.3±0.9, 85.±0.9%; $P<0.001$). SA1 ARL vocalized more than SA1 MRL during open field (60.8±10.9, 32.8±5.4; $P<0.05$) and social motivation tests (32.1±12.3, 10.4±3.9 respectively; $P<0.01$). Females vocalized more than males during open field (43.8±5.1, 27.0±3.2; $P<0.05$) and novel object tests (32.7±5.3, 13.1±2.3; $P<0.01$), interacted more frequently with the stimulus (0.29±0.05, 0.13±0.03; $P<0.05$), and spent more time close to other lambs (80.0±0.5, 66.1±0.7%; $P<0.001$), but vocalized less than males (7.7±2.1, 27.1±7.4; $P<0.01$) during social motivation tests. Overall, SA1 ARL were more fearful and socially motivated than SA1 MRL, suggesting either that detrimental effects of early maternal separation may be exacerbated by reduced space allowance during pregnancy, or that remaining with the mother buffers the detrimental effects of prenatal stress. Females were more curious towards novel object, but they overall acted more fearfully and searched for more social contact than males.

Authors index

Printed in the United States
by Baker & Taylor Publisher Services